Advances in Experimental Medicine and Biology

Neuroscience and Respiration

Volume 1222

Series Editor
Mieczyslaw Pokorski
Opole Medical School
Opole, Poland

More information about this series at http://www.springer.com/series/13457

Mieczyslaw Pokorski
Editor

Pulmonology

 Springer

Editor
Mieczyslaw Pokorski
Opole Medical School
Opole, Poland

ISSN 0065-2598   ISSN 2214-8019   (electronic)
Advances in Experimental Medicine and Biology
ISBN 978-3-030-34653-9   ISBN 978-3-030-34651-5   (eBook)
https://doi.org/10.1007/978-3-030-34651-5

This Springer imprint is published by the registered company Springer Nature Switzerland AG.
The registered company address is: Gewerbestrasse 11, 6330 Cham, Switzerland

Preface

The book series Neuroscience and Respiration presents contributions by expert researchers and clinicians in medical research and clinical practice. Particular attention is focused on pulmonary disorders as the respiratory tract is at the first line of defense against pathogens and environmental toxic sources. The articles provide timely overviews of contentious issues or recent advances in the diagnosis, classification, and treatment of the entire range of diseases and disorders, both acute and chronic. The texts are thought as a merger of basic and clinical research dealing with biomedicine at both molecular and functional levels and with the interactive relationship between respiration and other neurobiological systems, such as cardiovascular function, immunogenicity, endocrinology and humoral regulation, and the mind-to-body connection. The authors focus on modern diagnostic techniques and leading-edge therapeutic concepts, methodologies, and innovative treatments. Neuromolecular and carcinogenetic aspects relating to gene polymorphism and epigenesis as well as practical, data-driven options to manage patients also are addressed.

Body functions, including lung ventilation and its regulation, are ultimately driven by the brain. However, neuropsychological aspects of disorders are still mostly a matter of conjecture. After decades of misunderstanding and neglect, emotions have been rediscovered as a powerful modifier or even the probable cause of various somatic disorders. Today, the link between stress and health is undeniable. Scientists accept a powerful psychological connection that can directly affect our quality of life and health span.

Clinical advances stemming from molecular and biochemical research are but possible if research findings are translated into diagnostic tools, therapeutic procedures, and education, effectively reaching physicians and patients. All this cannot be achieved without a multidisciplinary, collaborative, bench-to-bedside approach involving both researchers and clinicians. The role of science in shaping medical knowledge and transforming it into practical care is undeniable.

Concerning respiratory disorders, their societal and economic burden has been on the rise worldwide, leading to disabilities and shortening of life-span. Chronic obstructive pulmonary disease and sleep apnea syndrome are cases in point. Concerted efforts are required to improve this situation, and part of those efforts is gaining insights into the underlying mechanisms of disease and staying abreast with the latest developments in diagnosis and treatment regimens. It is hoped that the articles published in this series will assume a leading position as a source of information on interdisciplinary medical research advancements, addressing the needs of medical professionals and allied health-care workers, and become a source of reference and inspiration for future research ideas.

I would like to express my deep gratitude to Paul Roos, and Cynthia Kroonen of Springer Nature NL for their genuine interest in making this scientific endeavor come through and in the expert management of the production of this novel book series.

Mieczyslaw Pokorski

Contents

Advs Exp. Medicine, Biology - Neuroscience and Respiration (2019) 44: 1–8
https://doi.org/10.1007/5584_2019_432
© Springer Nature Switzerland AG 2019
Published online: 21 September 2019

Spirometry: A Need for Periodic Updates of National Reference Values

Andrzej Chciałowski and Tomasz Gólczewski

Abstract

The aim of the study was to assess the need for changes in spirometry reference values in the Polish population with time lapse, as the after-effect of a radical socioeconomic overturn of the 1990. We retrospectively analyzed data files on forced expiratory volume in 1 s (FEV1), vital capacity (VC), and forced VC (FVC) in healthy, never-smoking Caucasians (731 females and 327 males) obtained in in 1993–1998. We assessed a discrepancy between the then measured values of these variables, on the one side, and the corresponding European Community for Steel and Coal (ECSC) predicted values or the current updated predicted values for the Polish population, on the other side. We found that those old measured values approximately corresponded to the ECSC reference, but they were appreciably lower than the current Polish reference values; the younger the subjects the greater the difference. The current Polish reference values of FVC were much closer to the old measured VC than to the old measured FVC values, which introduces a substantial discrepancy between the past and present FVCs. We conclude that the spirometry reference values may change with time lapse. Thus, accuracy of prediction equations should be periodically updated, which seems to particularly concern the equations elaborated for the nations that undergo rapid economic developments connected with changes in living standards.

Keywords

Forced expiration · Life expectancy · Lung function · Polish population · Prediction equations · Reference values · Socioeconomic conditions · Spirometry

1 Introduction

The diagnosis of obstructive lung diseases is founded on the comparison of spirometric measurements with reference values calculated for an individual by means of some prediction equations. Forced expiratory volume in 1 s (FEV1), forced vital capacity (FVC), and their ratio (FEV1/FVC) belong to the most important spirometric indices. To make a diagnosis these variables are compared with norms calculated for patient's age and height by means of some prediction equations. In the twentieth century, the European Community for Steel and Coal (ECSC) equations were the most often employed for

A. Chciałowski
Department of Infectious Diseases and Allergology, Military Institute of Medicine, Warsaw, Poland

T. Gólczewski (✉)
Laboratory of Cardiopulmonary System Diagnosis and Therapy Support, The Nalecz Institute of Biocybernetics and Biomedical Engineering, Polish Academy of Sciences, Warsaw, Poland
e-mail: tgolczewski@ibib.waw.pl

Caucasian populations (Quanjer et al. 1993). These equations, published in the final form in 1993, were based on the data collected for Western populations in the 1960–70s. However, almost all studies published since 1993 show that the ECSC equations, are not entirely accurate also for the present Polish population (Gólczewski 2012). In particular, it is clear that the plots of FEV1 against FVC lie far from those created on the basis of the most current equations (Gólczewski et al. 2012). Today, the Global Lung Function equations (GLI) (Quanjer et al. 2012) are suggested to be more appropriate in almost all ethnic groups. Nevertheless, recent works show some imperfections also in these equations (Kainu et al. 2016; Backman et al. 2015; Pereira et al. 2014; Ben Saad et al. 2013). Due to a clinical significance of the prediction equations, it seems warranted to study the underlying reasons of their declining accuracy.

It has been shown that socioeconomic status and living conditions of individuals may influence lung function (Rębacz-Maron et al. 2018; Taylor-Robinson et al. 2014; Hegewald and Crapo 2007; Choudhury et al. 1997). The equations elaborated previously also could lose the accuracy because of new guidelines for spirometry or due to changes in biological status of individuals related to alterations in lifestyle or living conditions. Life expectancy at birth, a commonly used indicator of health state (Gulis 2000), has to do with the political situation in a country, which, to some extent, is a determinant of living conditions (Mackenbach 2013). Radical political changes in Eastern Europe about three decades ago have led to rapid economic improvements, which provides an opportunity to investigate the influence of living conditions on lung function in a national population, with other variables held relatively constant. The reasons outlined above, related to the passage of time and changing conditions, were considered in this article. The old fundamental ECSC equations have been most known and are still in use in many clinical settings, despite the commonly accepted shortcomings. Therefore, the specific aim of this study was to compare FEV1 and FVC values measured in the past in a sample of the Polish population with predicted values using the old ECSC equations (Quanjer et al. 1993), on the one side, and the predicted values using the current Polish (PL) equations elaborated by Lubiński and Gólczewski (2010), on the other side. The study question was whether the correspondence of old measurements to the ECSC predicted values would be akin to that of the current PL predicted values. Any misalignment would be assumed to speak for changes with time lapse.

2   Methods

This is a retrospective study based on the dataset run by the Military Institute of Medicine in Warsaw, Poland, which contains results of pulmonary function tests conducted in a mobile laboratory in 180 Polish cities, towns, and villages during a 6-year long period from 1993 to 1998; a part of the then nationwide study on the evaluation of effects of smoking and air pollution on lung function. The data reviewed concerned 3878 women and 3362 men and consisted of the following parameters: age, height, weight, pack-years of smoking, exposure to air pollution, percentage of blood carboxyhemoglobin (HbCO %) verifying both exposure to pollution and smoking, and a history of respiratory symptoms such as cough, sputum production, and physical exertion tolerance. Subjects younger than 25 years of age, smokers or those with HbCO% >6%, or any of the symptoms outlined above were excluded from the analysis. After the exclusions, the evaluation was performed on the data of never-smoking individuals, numbering 731 females and 327 males (Table 1). Such a big gender difference stemmed from the predominance of male smokers in the time past. The evaluation was limited to the most essential spirometry variables of FEV1, FVC, and VC. The evaluation was performed in three age groups of 25–44, 45–59, and 60–80 years to take into account differences in living conditions and lifestyle possibly related to the conditions in which a particular generation had grown up.

Data were presented as means ±SD. The Shapiro–Wilk test was used to assess the normality of data distribution. Statistical differences

Table 1 Characteristics of the cohort investigated

	Females	Males
Number of subjects in database (n)		
Total	3878	3362
Excluded:		
ever-smokers	1696	2335
with pathological symptoms	1317	521
with non-medical reasons	53	81
aged <25 years	81	98
Remained (n)	731	327
Age (years)	52 ± 14 (25–85)	50 ± 15 (25–87)
Height (cm)	162 ± 6 (136–180)	174 ± 7 (151–192)
Basic spirometry variables		
FEV1 (L)	2.40 ± 0.57 (0.77–4.80)	3.46 ± 0.79 (1.40–5.60)
FVC (L)	2.82 ± 0.65 (1.23–5.33)	4.16 ± 0.92 (1.82–6.37)
VC (L)	3.06 ± 0.67 (1.23–5.33)	4.46 ± 0.93 (2.05–6.77)

Data are numbers or means ±SD with min-max range in parentheses

were assessed with one-sample t-test, with a null hypothesis that the mean difference between the value of a true test and the value for comparison would be equal to zero. A p-value <0.05 defined significant differences. The evaluation was performed using a commercial Statistica v10 package (StatSoft Inc., Tulsa, OK).

3   Results

Demographic details and basic values of spirometry variables in the population cohort under consideration are shown in Table 1.

Differences between the old measured FEV1 and FVC, on the one side, and the old ECSC predicted or the current PL predicted values, on the other side, had approximately normal distributions. The difference between the two respective sets of values (each set = measured minus predicted value) is graphically visualized in Fig. 1. This difference concerning FVC amounted to the considerable 0.4 L. Further, differences were comparable in magnitude in both female and male subjects. No such differences could be substantiated in case of FEV1, although the tendency was alike. Nor were there any appreciable differences concerning VC values (graphical data not shown).

Table 2 summarizes the mean differences, distributed by age, between the old measured and predicted values according to the ECSC equations versus the old measured and the current PL equations. Graphical age-related differences also are displayed in Fig. 2. Note that:

- FEV1 – there were no appreciable differences between the old measured and ECSC predicted values, whereas the old measured were significantly lower than the current PL predicted values; the younger the individuals the greater the difference;
- FVC – pattern akin to that for FEV1, with distinctly greater differences between the old measured and the current PL predicted values; particularly in case of younger age groups.

4   Discussion

Figure 2 shows that FVC and FEV1 values in males were slightly underestimated by ECSC equations in the upper range and overestimated in the lower range of these variables. Irrespective of whether the wrong estimations were due to some inherent imperfections in ECSC equations for males or by peculiar properties of exhibited by

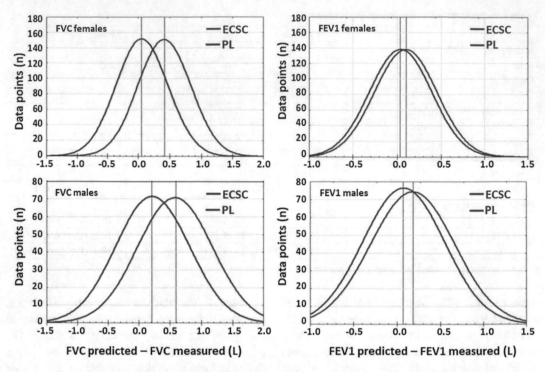

Fig. 1 Distribution of differences in forced expired volume in 1 s (FEV1) and forced vital capacity (FVC) between the old measured and predicted values according to the European Community for Steel and Coal equations (ECSC) (Quanjer et al. 1993) and the old measured and predicted values according to the current equations for the Polish population (PL) (Lubiński and Gólczewski 2010)

Table 2 Mean differences, distributed by age, between the old measured and predicted values of spirometry variables using the old European Community for Steel and Coal (ECSC) equations versus the old measured and predicted values using the current equations for the Polish population (PL)

	Females			Males		
Age group (year)	25–44	45–59	60–80	25–44	45–59	60–80
	n = 226	n = 244	n = 261 (179)	n = 127	n = 104	n = 96 (69)
FEV1 (L)						
ECSC	0.03	−0.04	−0.03	−0.02	−0.07	−0.10
PL	−0.17*	−0.10*	−0.02	−0.31*	−0.16*	−0.01
FVC (L)						
ECSC	−0.02	−0.05	−0.05	−0.09	−0.23*	−0.26*
PL	−0.50*	−0.43*	−0.30*	−0.73	−0.56*	−0.36*
VC versus FVC[a]						
PL	−0.26*	−0.19*	−0.05	−0.41*	−0.29*	−0.04

FEV1 forced expired volume in 1 s, *FVC* forced vital capacity, *VC* vital capacity
*p < 0.05; ECSC data were limited to subjects younger than 71 years of age whose numbers are in parentheses
[a]FVC according to current PL data of Lubiński and Gólczewski (2010)

Polish men during the past years, the ECSC equations for FEV1 and FVC were approximately appropriate for the Polish population at the time, since the then measured values did not appreciably differ from predictions, except FVC in older males (Table 2). Thus, a paradox appeared that in the last decade of the twentieth century:

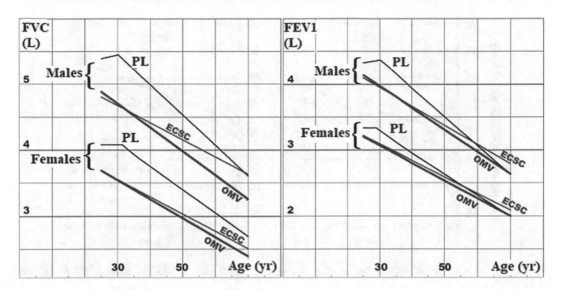

Fig. 2 FVC and FEV1 in female and male subjects aged over 25 years (162 cm and 170 cm of height, respectively) according to the old European Community for Steel and Coal (ECSC) equations (Quanjer et al. 1993) (red) and the current equations for the Polish population (PL) (Lubiński and Gólczewski 2010) (black). For comparison, the interpolated mean old measured values for such subjects also are shown (OMV, blue)

- ECSC equations were grossly adequate for the Polish Caucasian population despite that the elaboration of those equations was neither based nor took into account any of the Polish lung function measurements;
- ECSC equations were inadequate for the western Caucasian populations despite that the elaboration of those equations was based on the lung function data derived from those populations.

This paradox disappears if the physiological status of a national population, quantified by means of life expectancy at birth, is taken into account (Fig. 3). The life expectancy of Polish males in 1996 grossly corresponded to that of Western European males in 1964–1974 (Socio-Demographic Indicators 2018). Therefore:

- ECSC equations could be appropriate for the Polish population in the 1990s because Poles at that time physiologically corresponded to Western populations at the time when Western population data were tallied and elaborated for the lung function prediction equations;

- ECSC equations could not be appropriate for western populations in the 1990s and later years due to continual socioeconomic development of those populations.

The above reasoning is based on two assumptions: (1) lung function has to do with the physiological state of an individual, determined by living condition and (2) life expectancy at birth is a good index of the physiological state in relation to a whole nation. The first assumption related to individual subjects has been demonstrated, to some extent, in a number of studies as outlined in the introduction. The second assumption can also be accepted since life expectancy is a commonly used index of health status of a nation (Steel et al. 2018). These assumptions can help explain the following results:

- old measured values of lung function were smaller than the ECSC predictions in older individuals, particularly males, since life expectancy at birth shortly after the World War II was much lower in Poland than that in

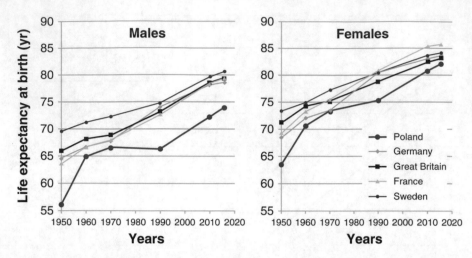

Fig. 3 Life expectancy at birth during passing decades in Poland and in Western Europe. (Compiled from Mackenbach's data of 2013)

Western Europe, which was due likely to the traumatic war-related events and a country destruction;

- large differences in younger subgroups of individuals between the old measured values of lung function and the current PL predicted values for FEV1 (Fig. 2) might be due to recent nationwide improvements in living conditions. The lack of such differences in the oldest individuals lend support for the notion that living conditions have a role in shaping lung function during the maturation time (Rębacz-Maron et al. 2018).

- difference between the old measured values of lung function and the current PL predicted values were smaller in females (Table 2) since they possibly were less sensitive to bad living conditions during the maturation time (Figs. 2 and 3).

Differences between the old measure values and the current PL predicted values for FVC were appreciably greater bigger than those for FEV1, and thus they may be underlain by the mechanisms other than just improved living conditions of individuals. If the expiration during lung testing is truly forced, FEV1 depends entirely on the physiological properties of the respiratory system linked to the airflow limitation. FVC also depends on these properties, but

contains an additional key element which is the duration of expiration. The current requirement for the duration of forced expiratory time ≥ 6 s (Miller et al. 2005), which was not in place at the time of creation of ECSC equations, is a likely cause of the greater differences in FVC noticed. When expiration is forcefully continued, a part of the air that is trapped at low lung volumes because of airflow limitation is slowly exhaled, which makes the current FVCs greater than old ones were. This seems supported by the fact that the current FVC is much closer to the past VC than to the past FVC. In all age groups, differences between the current predicted and measured FVC values were greater than those between the predicted and measured VC; by about 240 mL in females and 300 mL in males (Table 2). In consequence, the shortcomings of ECSC equations for FVC have to do not only with the possible neglect of improvements in living conditions, but also with the elongation of forced expiration beyond the 6 s. As noted by Hankinson et al. (2015), the past criterion for the end of forced expiration, i.e., a drop in airflow below 25 mL/s, is attained before the 6 s in individuals younger than 38 years of age, which might explain why the difference between the current measured and predicted FVC values in young subjects is much greater than that in older individuals.

The influence of socioeconomic status on lung function has been previously demonstrated, but the studies are cross-sectional and usually related to specific population groups, having various educational, occupational, or economic criteria, which are neither biological nor physiological. Therefore, a causal relationship between socioeconomic status and lung function is uncertain, as is also whether better status implies better lung function or the other way around. Nor is the role of other factors such as ethnicity or genotype, which may variably influence both socioeconomic status and lung function, well defined. In contradistinction, the present study is a kind of longitudinal study, concerning a whole ethnically uniform national population. Since lung function changes appeared in a rather short time of two decades or so, any evolutionary alterations could be hardly responsible either. Therefore, the findings seem cognizant with the influence of living conditions on lung function, particularly affecting younger individuals.

A major limitation of this study is the use of a rather old database; thereby some details related to procedures and equipment employed used at the time were different from the contemporary standard. In addition, spirometry tests performed before 2005 did not consider the forced expiration maneuver lasting for ≥ 6 s, which is a currently accepted standard according to the ATS/ERS guidelines (Miller et al. 2005). This drawbacks ought to be taken into account in the interpretation of the present findings. However, the old database was collated on the basis of the then valid ECSC predictions, which enables a verifiable integration and comparison with the data presently obtained.

We conclude that differences in the spirometry FEV1 test results taken between the past ECSC and current Polish recommendations, particularly evident in young male individuals, are caused, in all likelihood, by improvements in socioeconomic and living conditions during the last two decades. The ECSC prediction equations were adequate for the Polish population in the 1990s, since at that time it corresponded to the Western European population of 2–3 decades back in terms of living conditions. Those equations became inadequate for the Polish population in the twenty-first century due to rapid socioeconomic improvements, positively affecting biophysiological status. It appears that the universal national norms for lung function may be influenced by changing living standards and thus they are hardly stable and require periodic checks.

Conflicts of Interest The authors declare no conflicts of interest in relation to this article.

Ethical Approval All procedures performed in studies involving human participants were in accordance with the ethical standards of the institutional and/or national research committee and with the 1964 Helsinki declaration and its later amendments or comparable ethical standards. This study was approved by the Ethics Committee of the Military Institute of Medicine in Warsaw, Poland (permit no. 42/WIM/2016).

Informed Consent This is a retrospective study based on review of anonymous spirometry databases, which waives the requirement of obtaining consent from individual participants.

References

Backman H, Lindberg A, Sovijärvi A, Larsson K, Lundbäck B, Rönmark E (2015) Evaluation of the global lung function initiative 2012 reference values for spirometry in a Swedish population sample. BMC Pulm Med 15:26

Ben Saad H, El Attar MN, Hadj Mabrouk K, Ben Abdelaziz AB, Abdelghani A, Bousarssar M, Limam K, Maatoug C, Bouslah H, Charrada A, Rouatbi S (2013) The recent multi-ethnic global lung initiative 2012 (GLI2012) reference values don't reflect contemporary adult's North African spirometry. Respir Med 107:2000–2008

Choudhury S, Alam MS, Begum QN (1997) Lung function parameters of Bangladeshi male subjects in different living conditions. Bangladesh Med Res Counc Bull 23:30–33

Gólczewski T (2012) Spirometry: a comparison of prediction equations proposed by Lubiński for the Polish population with those proposed by the ECSC/ERS and by Falaschetti et al. Pneumonol Alergol Pol 80:29–40

Gólczewski T, Lubiński W, Chciałowski A (2012) A mathematical reason for FEV1/FVC dependence on age. Respir Res 13:57

Gulis G (2000) Life expectancy as an indicator of environmental health. Eur J Epidemiol 16(2):161–165

Hankinson JL, Eschenbacher B, Townsend M, Stocks J, Quanjer PH (2015) Use of forced vital capacity and forced expiratory volume in 1 second quality criteria for determining a valid test. Eur Respir J 45:1283–1292

Hegewald MJ, Crapo RO (2007) Socioeconomic status and lung function. Chest 132:1608–1614

Kainu A, Timonen KL, Toikka J, Qaiser B, Pitkaniemi J, Kotaniemi JT, Linqvist A, Vanninen E, Länsimies E, Sovijärvi ARA (2016) Reference values of spirometry for Finnish adults. Clin Physiol Funct Imaging 36:346–358

Lubiński W, Gólczewski T (2010) Physiologically interpretable prediction equations for spirometric indices. J Appl Physiol 108:1440–1446

Mackenbach JP (2013) Political conditions and life expectancy in Europe, 1900–2008. Soc Sci Med 82:134–146

Miller MR, Hankinson J, Brusasco V, Burgos F, Casaburi R, Coates A, Crapo R, Enright P, van der Grinten CPM, Gustafsson P, Jensen R, Johnson DC, MacIntyre N, McKay R, Navajas D, Pedersen OF, Pellegrino R, Viegi G, Wanger J (2005) Standardisation of spirometry. Eur Respir J 26:319–338

Pereira CAC, Duarte AAO, Gimenez A, Soares MR (2014) Comparison between reference values for FVC, FEV1, and FEV1/FVC ratio in white adults in Brazil and those suggested by the Global Lung Function Initiative 2012. J Bras Pneumol 40:397–402

Quanjer PH, Tammeling GJ, Cotes JE, Pedersen OF, Peslin R, Yernault JC (1993) Lung volumes and forced ventilatory flows. Report working party standardization of lung function tests, European Community for Steel and Coal. Official Statement of the European Respiratory Society. Eur Respir J 6(Suppl 16):S5–S40

Quanjer PH, Stanojevic S, Cole TJ, Baur X, Hall GL, Culver BH, Enright PL, Hankinson JL, Ip MSM, Zheng J, Stocks J (2012) The ERS Global Lung Function Initiative: multi-ethnic reference values for spirometry for the 3–95-yr age range: the global lung function 2012 equations. Eur Respir J 40:1324–1343

Rębacz-Maron E, Stangret A, Teul I (2018) Influence of socio-economic status on lung function in male adolescents in Tanzania. Adv Exp Med Biol 1150:53–67

Socio-Demographic Indicators (2018). http://countryeconomy.com/demography/life-expectancy. Accessed on Aug 31, 2018

Steel N, Ford JA, Newton JN, Davis ACJ, Vos T, Naghavi M, Glenn S, Hughes A, Dalton AM, Stockton D, Humphreys C, Dallat M, Schmidt J, Flowers J, Fox S, Abubakar I, Aldridge RW, Baker A, Brayne C, Brugha T, Capewell S, Car J, Cooper C, Ezzati M, Fitzpatrick J, Greaves F, Hay R, Hay S, Kee F, Larson HJ, Lyons RA, Majeed A, McKee M, Rawaf S, Rutter H, Saxena S, Sheikh A, Smeeth L, Viner RM, Vollset SE, Williams HC, Wolfe C, Woolf A, Murray CJL (2018) Changes in health in the countries of the UK and 150 English Local Authority areas 1990–2016: a systematic analysis for the Global Burden of Disease Study 2016. Lancet 392(10158):1647–1661

Taylor-Robinson DC, Thielen K, Pressler T, Olesen HV, Diderichsen F, Diggle PJ, Smyth R, Whitehead M (2014) Low socioeconomic status is associated with worse lung function in the Danish cystic fibrosis population. Eur Respir J 44:1363–1366

Advs Exp. Medicine, Biology - Neuroscience and Respiration (2019) 44: 9–16
https://doi.org/10.1007/5584_2019_415
© Springer Nature Switzerland AG 2019
Published online: 19 July 2019

Age at Menarche and Risk of Respiratory Diseases

Martyna Zurawiecka and Iwona Wronka

Abstract

Several studies have suggested a relationship between reproductive history and respiratory health. The present study explores the association between the age at menarche and the risk of respiratory diseases during early adulthood. The anthropometrical and questionnaire research was conducted among 1323 female university graduates. In a subsample of 152 non-allergic women spirometry tests were additionally performed. We found that the prevalence of allergic diseases, on average, was slightly higher among females having early menarche than in those with a later onset of menstruation; the difference failed to reach statistical significance. However, the risk of allergic rhinitis was significantly related with early menarche compared with average-time, taken as a reference, or late menarche (OR = 1.61 *vs*. OR = 1 (Ref.) *vs*. OR = 1.23, p = 0.020). The difference remained significant after adjusting for adiposity (p = 0.050) and socio-economic status (p = 0.001). There was no significant relationship between the age at menarche and the incidence of respiratory infections. We noticed a tendency for increased spirometry variables with increasing age at menarche. In conclusion, early menarche is a risk factor for allergic rhinitis in early adulthood.

Keywords

Age · Allergic rhinitis · Early menarche · Respiratory infection · Spirometry

1 Introduction

Menarche is the first menstrual bleeding and represents the beginning of reproductive life. It is a key developmental marker of a girl's healthy transition from childhood into young adulthood. The age at menarche has been evolving for much of the human history. Since the middle of the twentieth century, the tendency for the puberty to occur earlier in consecutive birth cohorts has been observed in the majority of developed as well as developing countries. Nowadays, more and more girls experience their first menstruation at a very young age, i.e., below 11 years (Adams Hillard 2008). A number of studies have analyzed the causes and results of this phenomenon (Bell et al. 2018; Gill et al. 2017; Macsali et al. 2011).

The age at menarche reflects the interaction between genotype and environment. Key external factors reported as having an effect on the age at menarche include the socio-economic status, which can be used to an indirect determination of the lifestyle and physical activity in childhood, health in childhood, psychological aspects, and the exposure to environmental pollutants

M. Zurawiecka (✉) and I. Wronka
Department of Anthropology, Institute of Zoology and Biomedical Research, Faculty of Biology, Jagiellonian University, Cracow, Poland
e-mail: martyna.zurawiecka@doctoral.uj.edu.pl

(Karapanou and Papadimitriou 2010). Substances present in the ambient air, food, and everyday supplies, affecting the hormonal system, are of notable importance (Fisher and Eugster 2014).

The age at menarche is considered a reliable marker of sex hormones, as it is related not only to the length of exposure to estrogens, but also to their level. Early menarche females have a higher level of estrogen and a lower one of serum hormone binding globulin. Overall, they are exposed to greater cumulative doses of estrogen and progesterone at adulthood than late menarche ones (Apter et al. 1989). Long-term exposure to high estrogen levels has an adverse impact on health. Thus, early menarche facilitates the occurrence of obesity, diabetes type 2, and it also affects cardiovascular function (Bell et al. 2018; He et al. 2010; Aksglaede et al. 2009; Lakshman et al. 2009; Remsberg et al. 2005). In addition, early menarche is a risk factor of breast cancer and ovarian cancer (Gong et al. 2013).

Research over the past few years have also linked the puberty acceleration to respiratory health (Macsali et al. 2012). Sex hormones have been shown to influence the inflammatory processes in lungs (Salam et al. 2006), smooth muscles (Herrera-Trujillo et al. 2005), and also the immune system (Xu et al. 2000). Estrogen affects the inflammatory mediators, T cell population, and the production of specific antibodies (Almqvist et al. 2008). Moreover, it may also trigger the growth of bronchial smooth muscles, which affects pulmonary contractility (Jain et al. 2012; Haggerty et al. 2003). Early menarche, associated with high estrogen levels, may thus lead to respiratory disorders. The findings reported to-date suggest that early menarche is related to the prevalence of asthma (Al-Sahab et al. 2010; Varraso et al. 2005) and allergic diseases (Klis and Wronka 2017; Westergaard et al. 2003; Xu et al. 2000) and it affects spirometry variables (Macsali et al. 2011). However, research on the subject is rather scant and not fully convincing. Further studies seem warranted to assess various aspects of menarcheal age and the propensity for respiratory disorders at later life periods.

The purpose of this study was to evaluate the relationship between the age at menarche and the prevalence of allergic diseases, with particular emphasis on allergic rhinitis, and to verify whether early menarche is a risk factor for a frequent occurrence of upper respiratory tract infections. Since both menarcheal age and respiratory health are connected with the level of adiposity and living conditions, we took into account the socioeconomic status and anthropometric indicators of general and abdominal obesity. In addition, we sought to evaluate the relationship between the age at menarche and lung function in a subgroup of healthy female subjects.

2 Methods

2.1 Participants, Questionnaires, and Measurements

Data were collected from 1323 female university graduates aged 19–25; the mean age of 20.6 ± 1.4SD years. All participants were Caucasian. Females (up to the age of full 25 years minus 1 day) who were non-pregnant, childless, and not currently suffering from any disorder were selected for inclusion into the study. The were no differences in the mean age between groups divided into categories, according to puberty age, prevalence of allergy, and prevalence of allergic rhinitis. Women older than 25 years and those who did not meet the inclusion criteria were excluded.

Questions contained in the survey were related to subjects' age at the first menstruation, the presence of allergy, allergic rhinitis, and the frequency of upper respiratory tract infections. The subjects were divided into three categories according to percentile values. The boundary value of the 25th and 75th percentiles (12–14 years) was considered the average age at menarche. Females whose menarche occurred before the age of 12 (25th percentile) were classified as early-maturing, while those whose menarche occurred after the age of 14 (75th percentile) were classified as late-maturing.

Questions concerning the overall health were asked to obtain the information about the prevalence of chronic conditions, including allergic conditions, allergic rhinitis, and the frequency of infectious respiratory diseases. The prevalence of allergies and/or allergic rhinitis was determined using the response to the question "Have you been diagnosed with allergy on the basis of medical tests, and if yes, which allergens are you allergic to?" A question on the presence of the symptoms of allergic rhinitis was also asked, i.e., about sneezing, running or blocked nose, sometimes with itchy eyes or nose. The reason behind this question was to verify whether there would have been any undiagnosed cases of allergic rhinitis among individuals who had never undergone relevant medical tests. None of such individuals declared the presence of hay fever symptoms.

The amount and distribution of adiposity and socio-economic status were selected as covariates in the analysis. The anthropometric indicators consisted of body height, body mass, chest, and waist circumference. Body height was measured with accuracy to 0.5 cm and weight in light cloths with accuracy to 0.5 kg. The waist and chest circumference were measured with an anthropometric tape. Waist circumference was measured halfway between the lower rib margin and the iliac crest. Chest circumference was measured at the xiphoidale point, located on the body of the sternum. The following parameters were calculated:

- body mass index (BMI; kg/m^2)
- waist-to-height ratio (WHtR; cm/cm)
- waist-to-chest ratio (WCR; cm/cm)

According to the guidelines of World Health Organization (WHO 2008), BMI is used to determine the prevalence of obesity and abdominal obesity. Subjects with a BMI below 18.5 kg/m^2 are qualified as underweight, with 18.5–25 kg/m^2 as of normal body mass, with 25–30 kg/m^2 as overweight, and > 30 kg/m^2 as obese. A WHtR below 0.4 indicates an insufficient adiposity level

in the abdominal region, 0.4–0.5 – normal level, 0.5–0.6 – heightened level, and values >0.6 indicate abdominal obesity. The WCR was analyzed as a continuous variable.

Socioeconomic status took into account the place of residence, parents' level of education, and the number of siblings. The place of residence was considered in the categories of village, town (up to 100,000 inhabitants), and city (more than 100,000 inhabitants). Parents' education was categorized as vocational, secondary, and college/university. The number of siblings was top coded at four. Participants were asked to self-assess their economic stance during childhood and adolescence as 1 – poor, 2 – average, 3 –good, 4 – very good, and 0 – changeable or difficult to assess. As there were only three responses in the zero category, it was discarded in the final analysis. On the basis of the data above outlined, a complex indicator of socioeconomic status was calculated, stratifying the participants into the groups of low, average, and high status. This stratification was carried out on the basis of a value of the first component obtained in the principal component analysis (PCA 2018).

2.2 Statistical Elaboration

Data were shown as means ±SD. The relationships between categorical variables such as age at menarche and prevalence of diseases was examined by means of a Chi^2 test. The significance of differences in the mean age at menarche, depending on the occurrence of allergy and allergic rhinitis, were analyzed by Student's t-test. Logistic regression analysis was applied to evaluate the risk of allergic rhinitis depending on the age at menarche. MANOVA was used to evaluate the difference in spirometry variables between early-, average-, and late-maturing women. A p-value of <0.05 defined statistically significant differences. Calculations using a commercial Statistica v12 package (StatSoft; Tulsa, OK).

3   Results

3.1   Age at Menarche and Allergy

As many as 943 (71.2%) females participating in the study reported no allergies, and 380 (28.8%) reported a medically diagnosed allergy. In the distribution by age at menarche, we found no statistical differences in the prevalence of allergies among the three groups (p = 0.134). Nonetheless, the occurrence of allergy tended to be somehow higher in the early-maturing than in the average- and late-maturing females; 32.0% vs. 26.3% vs. 30.3%, respectively. Next, we analyzed the age at menarche specifically in connection with allergic rhinitis. Rhinitis was reported by 243 out of the 1,323 females. It occurred significantly more frequently in the females having early menarche than in those who experienced late menarche (23.2% vs. 15.8% vs. 18.8%, respectively) (p = 0.018). Infections of the upper airways were occasionally occurring in all the females investigated. The frequency of infection was rather modest. Females who suffered from the infection rarely, i.e., less than 3 per year, outnumbered about 15-fold those who had more than 3 infections per year; 1,239 vs. 84 females, respectively. There was no statistically significant difference in the number of upper airway infections by the age at menarche (Table 1).

Logistic regression was used to estimate the risk of allergic rhinitis depending on the age at menarche. In the group of early menarche, odds ratio (OR) was 1.61 with 95% confidence intervals (95%CI) of 1.16–2.24, whereas it was 1.23 (95%CI = 0.87–1.74) for the late mnenarche. Females with the average age at menarche constituted a group of reference for comparison (OR = 1.00). The inter-group differences were significant (p = 0.020) and remained so after adjustments for BMI (p = 0.030), WHtR (p = 0.030), or WCR (p = 0.021). After combining menarcheal age with the three variables BMI, WHtR, and WCR, pertaining to the amount and distribution of adipose tissue, the p-value remained borderline significant, equalling 0.05. The relationship between menarcheal age and likelihood of allergic rhinitis also remained significant after adjustment for socio-economic variables (p-values varied from 0.01 to 0.04).

Finally, we evaluated whether menarcheal age could influence pulmonary function. Spirometry tests were performed in 152 healthy females of this study, who were free of any chronic conditions, including the allergic ones. The analysis, with BMI and the prevalence of abdominal obesity as covariates, failed to demonstrate any statistically relevant differences in spirometry results, depending on the age at menarche. However, there was a tendency for increased FVC, FEV1, and FEV1/FVC ratio with increasing age at menarche (Table 2).

Table 1 Prevalence of allergy, allergic rhinitis, and upper respiratory tract infections in relation to age at menarche

		Age at menarche		
		Early (n = 328)	Average (n = 665)	Late (n = 330)
Allergy	No; n (%)	223 (68.0)	490 (73.7)	230 (69.7)
	Yes; n (%)	105 (32.0)	175 (26.3)	100 (30.3)
		Chi^2 = 4.02 p = 0.134		
Allergic rhinitis	No; n (%)	252 (76.8)	560 (84.2)	268 (81.2)
	Yes; n (%)	76 (23.2)	105 (15.8)	62 (18.8)
		Chi^2 = 8.03; p = 0.018		
Upper respiratory tract infection	≥3 times/year; n (%)	23 (7.0)	40 (6.0)	21 (6.4)
	1–2 times//year; n (%)	163 (49.7)	317 (47.7)	159 (48.2)
	Less often; n (%)	142 (43.3)	308 (46.3)	150 (45.4)
		Chi^2 = 0.68; p = 0.954		

Early age at menarche <12 years, average 12–14 years, and late >14 years

Table 2 Spirometry variables depending on age at menarche

| Variable | Age at menarche | | | p |
	Early (n = 22)	Average (n = 99)	Late (n = 28)	
FCV (L)	3.40 ± 0.55	3.48 ± 0.48	3.63 ± 0.65	0.445
FVC (% predicted)	0.92 ± 0.15	0.95 ± 0.12	0.98 ± 0.16	0.243
FEV_1 (L)	2.90 ± 0.46	3.02 ± 0.44	3.17 ± 0.49	0.092
FEV_1 (% predicted)	0.88 ± 0.14	0.93 ± 0.12	0.97 ± 0.14	0.087
PEF(L)	4.69 ± 1.56	4.45 ± 1.27	4.91 ± 1.21	0.098
PEF (% predicted)	0.66 ± 0.20	0.64 ± 0.18	0.69 ± 0.16	0.143
FEV_1/FVC ratio	0.86 ± 0.10	0.87 ± 0.12	0.88 ± 0.14	0.211

Data are means $\pm$SD. Early age at menarche <12 years, average 12–14 years, and late >14 years; comparison among the age at menarche groups with one-way ANOVA

4 Discussion

Initially, studies on the influence of sex hormones on respiratory health dealt with gender differences regarding the incidence of diseases. It has been reported that in childhood asthma is more common in males, whereas in adolescence and adulthood in females (Becklake and Kauffmann 1999; Skobeloff et al. 1992) or that women experience more severe asthma than men (Henriksen et al. 2003; de Marco et al. 2000). The influence of estrogens on respiratory function is confirmed by changes in spirometry and the more frequent occurrence of asthma attacks during the menstrual cycle, the relationship between the occurrence of respiratory and reproductive system disorders, e.g., irregular menstrual cycles or polycystic ovary syndrome, and during the use of contraceptives (Real et al. 2007; Salam et al. 2006; Svanes et al. 2005).

Menarche initiates regular menstrual cycles and marks a high level of sex hormones. It completes the process of growth and maturation. The younger the age at the first menstruation, the shorter is the stage of progressive development, and the longer is the period in which a female remains under the influence of high sex hormone levels. A shortened period of growth and development of the body, including the respiratory system, and extended exposure to estrogens have health consequences. There are reports that females who mature earlier more frequently suffer from asthma or display asthmatic symptoms

(Varraso et al. 2005; Guerra et al. 2004). According to a study by Al-Sahab et al. (2011) earlier menarche doubles the risk of asthma after puberty. A similar relationship has been observed by Salam et al. (2006) in a group of adult women. Early age at menarche also is reported to contribute to more frequent occurrence of asthma symptoms or to asthma with bronchial hyperreactivity. Such findings are consistent all over Europe, which puts in doubt a possible impact of socio-economic factors (Macsali et al. 2011).

Fewer studies address the relationship between the age at menarche and allergic conditions. In a study on 5,000 females, Xu et al. (2000) have observed that earlier menarche associates with a rise in the prevalence of atopy. Westergaard et al. (2003) have demonstrated a relationship between the age at menarche and allergic rhinitis in a study on 3,000 females. Girls who first menstruated at the age of 12 years are 1.25 times more likely to have allergic rhinitis than those in whom menarche occurred at 14 years. The results of the present study revealed a similar significant relationship between the age at menarche and allergic rhinitis. The likelihood of developing allergic rhinitis was adversely related to the age at menarche. Likewise, there was a consistent tendency for allergies, taken together, to occur more often in the subjects who reached puberty at younger age, although in this case the difference failed to reach statistical significance.

The mechanisms that underlie the impact of sex hormones on the development of allergic conditions have most likely to do with affecting the immune system function. The occurrence of allergic rhinitis at early menarche may also have to do with body mass. The evidence shows that females of early menarche usually have increased body mass (Bell et al. 2018; Prentice and Viner 2013). It has been demonstrated that obesity disrupts the function of both respiratory and immune systems (Vatrella et al. 2016; Lessard et al. 2011; Steele et al. 2009). Nevertheless, both previous studies and the present one demonstrate that the relationship between the age at menarche and respiratory health is also noticeable after adjustment for BMI (Macsali et al. 2011). This indicates that there could be a specific effect on respiratory health of high estrogen levels rather than that of adiposity. Interestingly, the reverse also holds true, i.e., obese individuals often have high estrogen levels.

The relationship between the age at menarche and respiratory allergies may also have to do with socio-economic factors. Females of high standard status often belong to those having both very early menarche and frequent occurrence of allergic diseases (Asher et al. 2006). Co-existence of the two, however, is not consistently reported in the literature. Nor did the present study show any appreciable connection of the age at menarche with allergies, depending on socio-economic status.

The evidence of the impact of early menarche on lung function also is limited. Macsali et al. (2011) have analyzed data collected from adult females and conclude that women who started to menstruate before the age of 11 had lower values of FEV_1 and FVC than those whose regular cycles had begun after the age of 13. However, these authors do not notice any relationship between early age at menarche and chronic obstructive pulmonary disease, due perhaps to the fact that the study group consisted of very young subjects. In the present study, spirometry was performed in a limited subgroup of 152 females. The results also indicate a tendency for lower FEV_1, FVC, and FEV_1/FVC in females with earlier menarche; a tendency that was close to borderline significance for FEV_1. It is worth noting that all these females were young and healthy, which could bear on the lack of significant differences.

Research discussed in the present article has certain limitations such as the self-reported study. Despite additional questions on medical tests and symptoms of allergic rhinitis one cannot be exclude that in the study group of non-allergic subjects were also individuals with undiagnosed allergic diseases. In addition, the study group comprised female university students, which means that some form of selection had already been applied. The subjects included individuals from families of varied status and different regions of Poland, and the tests were conducted in several academic centres: both at large universities and local colleges.

In summary, the results of our research show that early menarche is a risk factor for allergic rhinitis at early adulthood in this cohort. The findings also suggest that early menarche could act as a curb on lung function at later age.

Conflicts of Interest The authors declare no conflicts of interest in relation to this article.

Ethical Approval All procedures performed in studies involving human participants were in accordance with the ethical standards of the institutional and/or national research committee and with the 1964 Helsinki declaration and its later amendments or comparable ethical standards. The study was approved by the institutional Ethics Committee.

Informed Consent Informed consent was obtained from all individual participants included in the study.

References

Adams Hillard PJ (2008) Menstruation in adolescents: what's normal, what's not. Ann N Y Acad Sci 1135:29–35

Aksglaede L, Sorensen K, Petersen JH, Skakkebaek NE, Juul A (2009) Recent decline in age at breast development: the Copenhagen Puberty Study. Pediatrics 123:932–939

Almqvist C, Worm M, Leynaert B (2008) Impact of gender on asthma in childhood and adolescence: a GA2LEN review. Allergy 63:47–57

Al-Sahab B, Hamadeh MJ, Ardern CI, Tamim H (2011) Early menarche predicts incidence of asthma in early adulthood. Am J Epidemiol 173:64–70

Apter D, Reinila M, Vihko R (1989) Some endocrine characteristics of early menarche, a risk factor for breast cancer, are preserved into adulthood. Int J Cancer 44:783–787

Asher MI, Montefort S, Björkstén B, CKW L, Strachan DP, Weiland SK, Williams H, ISAAC Phase Three Study Group (2006) Worldwide time trends in the prevalence of symptoms of asthma, allergic rhinoconjunctivitis, and eczema in childhood: ISAAC Phases One and Three repeat multicountry cross-sectional surveys. Lancet 368:733–743

Becklake MR, Kauffmann F (1999) Gender differences in airway behaviour over the human life span. Thorax 54:1119–1138

Bell JA, Carslake D, Wade KH, Richmond RC, Langdon RJ, Vincent EE, Holmes MV, Timpson NJ, Smith DG (2018) Influence of puberty timing on adiposity and cardiometabolic traits: a Mendelian randomisation study. PLoS Med 15:e100264

de Marco R, Locatelli F, Sunyer J, Burney P (2000) Differences in incidence of reported asthma related to age in men and women. A retrospective analysis of the data of the European Respiratory Health survey. Am J Respir Crit Care Med 162:68–74

Fisher MM, Eugster EA (2014) What is in our environment that effects puberty? Reprod Toxicol 44:7–14

Gill D, Sheehan NA, Wielscher M (2017) Age at menarche and lung function: a Mendelian randomization study. Eur J Epidemiol 32:701–710

Gong TT, Wu QJ, Vogtmann E, Lin B, Wang YL (2013) Age at menarche and risk of ovarian cancer: a meta-analysis of epidemiological studies. Int J Cancer 132:2894–2900

Guerra S, Wright AL, Morgan WJ, Sherrill DL, Holberg CJ, Martinez FD (2004) Persistence of asthma symptoms during adolescence: role of obesity and age at the onset of puberty. Am J Respir Crit Care Med 170:78–85

Haggerty CL, Ness RB, Kelsey S, Waterer GW (2003) The impact of estrogen and progesterone on asthma. Ann Allergy Asthma Immunol 90:284–291

He C, Zhang C, Hunter DJ, Hankinson SE, Buck Louis GM, Hediger ML, Hu FB (2010) Age at menarche and risk of type 2 diabetes: results from 2 large prospective cohort studies. Am J Epidemiol 171:334–344

Henriksen AH, Holmen TL, Bjermer L (2003) Gender differences in asthma prevalence may depend on how asthma is defined. Respir Med 97:491–497

Herrera-Trujillo M, Barraza-Villarreal A, Lazcano-Ponce-E, Hernández B, Sanín LH, Romieu I (2005) Current wheezing, puberty, and obesity among Mexican adolescent females and young women. J Asthma 42:705–709

Jain R, Ray J, Pan J, Brody S (2012) Sex hormone-dependent regulation of cilia beat frequency in airway epithelium. Am J Respir Crit Care Med 46:446–453

Karapanou O, Papadimitriou A (2010) Determinants of menarche. Reprod Biol Endocrinol 8:115–123

Klis K, Wronka I (2017) Association of estrogen-related traits with allergic rhinitis. Adv Exp Med Biol 968:71–78

Lakshman R, Forouhi NG, Sharp SJ, Luben R, Bingham SA, Khaw KT, Wareham NJ, Ong KK (2009) Early age at menarche associated with cardiovascular disease and mortality. J Clin Endocrinol Metab 94:4953–4960

Lessard A, Alméras N, Turcotte H, Tremblay A, Despres JP, Boulet LP (2011) Adiposity and pulmonary function: relationship with body fat distribution and systemic inflammation. Clin Invest Med 34:64–70

Macsali F, Real FG, Plana E, Sunyer J, Anto J, Dratva J, Janson C, Jarvis D, Omenaas ER, Zemp E, Wjst M, Leynaert B, Svanes C (2011) Early age at menarche, lung function, and adult asthma. Am J Respir Crit Care Med 183:8–14

Macsali F, Svanes C, Bjorge L, Omenaas ER, Gomez RF (2012) Respiratory health in women: from menarche to menopause. Expert Rev Respir Med 6:187–200

PCA (2018) Principal components analysis. Institute for Digital Research and Education, UCLA. https://stats. idre.ucla.edu/sas/output/principal-components-analysis/; Accessed on 6 June 2019

Prentice P, Viner RM (2013) Pubertal timing and adult obesity and cardiometabolic risk in women and men: a systematic review and metaanalysis. Int J Obes 37:1036–1043

Real FG, Svanes C, Omenaas ER, Antò JM, Plana E, Janson C, Jarvis D, Zemp E, Wjst M, Leynaert B, Sunyer J (2007) Menstrual irregularity and asthma and lung function. J Allergy Clin Immunol 120:557–564

Remsberg KE, Demerath EW, Schubert CM, Chumlea WC, SunSS SRM (2005) Early menarche and the development of cardiovascular disease risk factors in adolescent girls: the FELS longitudinal study. J Clin Endocrinol Metab 90:2718–2724

Salam MT, Wenten M, Gilliland FD (2006) Endogenous and exogenous sex steroid hormones and asthma and wheeze in young women. J Allergy Clin Immunol 117:1001–1007

Skobeloff EM, Spivey WH, St Clair SS, Schoffstall JM (1992) The influence of age and sex on asthma admissions. JAMA 268:3437–3440

Steele RM, Finucane FM, Griffin SJ, Warcham NJ, Ekelund U (2009) Obesity is associated with altered lung function independently of physical activity and fitness. Obesity 17:578–584

Svanes C, Real FG, Gislason T, Jansson C, Jogi R, Norrman E, Nystrom L, Toren K, Omenaas E (2005) Association of asthma and hay fever with irregular menstruation. Thorax 60:445–450

Varraso R, Siroux V, Maccario J, Pin I, Kauffmann F (2005) Asthma severity is associated with body mass index and early menarche in women. Am J Respir Crit Care Med 171:334–339

Vatrella A, Calabrese C, Mattiello A, Panico C, Costigliola A, Chiodini P, Panico S (2016) Abdominal adiposity is an early marker of pulmonary function impairment: findings from a Mediterranean Italian female cohort. Nutr Metab Cardiovasc Dis 26:643–648

Westergaard T, Begtrup K, Rostgaard K, Krause TG, Benn CS, Melbye M (2003) Reproductive history and allergic rhinitis among 31145 Danish women. Clin Exp Allergy 33:301–305

WHO (2008) Waist circumference and wait to hip ratio. Report of a WHO expert consultation. World Health Organization, Geneva. https://www.who.int/nutrition/publications/obesity/WHO_report_waistcircumference_and_waisthip_ratio/en/; Accessed on 6 June 2019

Xu B, Jarvelin MR, Hartikainem AL, Pekkanen J (2000) Maternal age at menarche and atopy among offspring at the age of 31 years. Thorax 55:691–693

Advs Exp. Medicine, Biology - Neuroscience and Respiration (2019) 44: 17–25
https://doi.org/10.1007/5584_2019_418
© Springer Nature Switzerland AG 2019
Published online: 21 September 2019

Effects of Osteopathic Manual Therapy on Hyperinflation in Patients with Chronic Obstructive Pulmonary Disease: A Randomized Cross-Over Study

M. Maskey-Warzechowska, M. Mierzejewski, K. Gorska, R. Golowicz, L. Jesien, and R. Krenke

Abstract

Osteopathic manual therapy (OMT) may reduce hyperinflation in patients with chronic obstructive pulmonary disease (COPD) by improving breathing mechanics. The aim of the study was to evaluate the immediate effects of OMT on hyperinflation in stable COPD patients with forced expired volume in 1 s (FEV$_1$) <50% predicted. Nineteen COPD patients of the median age 68 (IQR 63–72) years and the median FEV$_1$ 39.8 (IQR 33.4–46.6) % predicted were enrolled into the study. For the first session, patients were randomly assigned to either OMT or sham therapy. During the second session, the two groups of patients were crossed over. Pulmonary function and dyspnea were compared before and after both procedures. Neither pulmonary function nor dyspnea differed significantly before and after OMT or sham procedures. However, 36.7% and 47.4% patients achieved the minimally important difference for residual volume (RV) reduction after both OMT and sham therapy, respectively. Responders to OMT had a greater median (IQR) baseline sense of dyspnea compared to non-responders, assessed on a visual analog scale, of 7.0 (4.5–7.0) vs. 3.0 (0.0–5.0), p = 0.040, respectively. Although OMT did not have an immediate effect on hyperinflation or dyspnea, a subgroup experienced a reduction in RV following OMT and sham therapy. Future studies are needed to identify the characteristics of responders.

Keywords

Airway obstruction · COPD · Dyspnea · Hyperinflation · Osteopathic therapy · Pulmonary function · Residual volume

M. Maskey-Warzechowska, M. Mierzejewski, K. Gorska (✉), and R. Krenke
Department of Internal Medicine, Pulmonary Diseases and Allergy, Medical University of Warsaw, Warsaw, Poland
e-mail: katarzyna.gorska@wum.edu.pl

R. Golowicz and L. Jesien
Flanders International College of Osteopathy, Warsaw, Poland

1 Introduction

The 2019 update of the Global Initiative for Chronic Obstructive Lung Disease (GOLD) highlights the role of disease symptoms and exacerbation risk as the two main determinants of disease progression and points to symptom alleviation and reduction of exacerbation frequency as important treatment targets (GOLD 2019). Although much progress has been done in the pharmacotherapy of chronic obstructive pulmonary disease (COPD) in the past decade, treatment efficacy is still far from satisfactory. Physiotherapeutic rehabilitation, which plays a key role in patients with stable COPD, focuses mainly on endurance training and muscle mass improvement (Celli 2017; Bolton et al. 2013). Little attention is given to chest wall mechanics *per se*. Respiratory muscle overload and increased wall rigidity due to an increased volume of air remaining in the lungs after spontaneous expiration in patients with airflow obstruction (static hyperinflation) lead to increased tension of the respiratory muscles and a decreased mobility of chest wall joints, what may further affect the work of breathing and the sense of dyspnea. There is evidence that impairment of pulmonary function in patients with COPD is associated with adaptive changes in the length and mobility of the muscles of the chest, thoracic spine and shoulders (Morais et al. 2016).

The concept that the chest wall could be a potential target for non-pharmacologic therapeutic intervention in patients with respiratory diseases dates back to the 1970s (Howell et al. 1975). Osteopathic manual therapy (OMT) is a method of manual therapy involving not only the soft tissues, but also bones and joints. The effectiveness of OMT has been analyzed in patients with moderate and severe COPD, with the assumption that it may improve breathing mechanics and reduce dyspnea related with increased airway resistance and increased residual volume (RV). An increase in RV results in a lower inspiratory capacity (IC) and thereby in a lower operating tidal volume what is considered one of the most important factors leading to dyspnea in hyperinflated COPD patients. Therefore, a reduction in RV is an important therapeutic target in COPD (Langer et al. 2014). Studies on the effect of OMT on RV and breathing mechanics produced conflicting results (Engel et al. 2013; Zanotti et al. 2012; Noll et al. 2008). Individual studies have shown a positive effect of manual therapy on chest wall mechanics, perception of dyspnea and peripheral oxygen saturation (SpO_2) in patients with COPD even after a single therapeutic session (Cruz–Montecinos et al. 2017; Yilmaz Yelvar et al. 2016). This has not been

unequivocally confirmed in systematic reviews (Steel et al. 2017; Cicchitti et al. 2015; Heneghan et al. 2012). On the other hand, Hogg et al. (2012) have shown that despite a documented beneficial effect of long-term pulmonary rehabilitation, patient adherence to therapy is below 50%. It seems that acute symptom alleviation after a single OMT session in patients with dyspnea is an attractive option for non-pharmacological intervention as it does not require the long-term compliance. Therefore, we undertook a randomized, cross-over study to evaluate the immediate effect of OMT on hyperinflation in patients with severe COPD.

2 Methods

2.1 Study Design

The study was performed between April and July 2016 and included COPD patients with severe airway obstruction (FEV1 < 50%). The effect of a single therapeutic session of OMT was compared to that of a sham procedure. Each patient underwent both OMT and sham procedure in a two-week interval. Before the first session the patients were randomized to OMT or sham procedure and after 2 weeks the patients were crossed-over (Fig. 1). Computer-generated random numbers were used to allocate patients to OMT or sham protocol during the first session. Lung function and severity of dyspnea were assessed directly before and after each session of OMT and sham procedure. The primary endpoint was the change between pre- and post-procedure RV, while the secondary endpoints were changes in the other indices of lung hyperinflation, such as total lung capacity (TLC), RV/TLC, functional residual capacity (FRC), FRC/TLC, inspiratory capacity (IC), and IC/TLC, and the perception of dyspnea assessed on a visual analog scale (VAS). A separate comparative analysis of patients who would demonstrate a post-intervention minimally important difference (MID) for RV and RV/TLC and those who would not was performed. We adopted the MID threshold of −310 mL and − 6.1% for RV and − 2.8% for RV/TLC based on a study of Hartman et al. (2012).

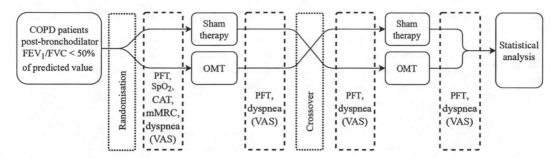

Fig. 1 Flow chart representing the study paradigm. *COPD* chronic obstructive pulmonary disease, *FEV₁* forced expiratory volume in 1 s, *FVC* forced vital capacity, *PFT* pulmonary function testing; *SpO₂* peripheral oxygen saturation measured by pulse oximetry, *CAT* COPD assessment test, *mMRC* modified Medical Research Council scale for dyspnea, *VAS* visual analogue scale, *OMT* osteopathic manual therapy

2.2 Study Patients

The study included stable patients with severe or very severe airway obstruction (FEV1 < 50%). The patients were recruited from the pulmonary outpatient department of the Public Central Teaching Hospital in Warsaw. Nineteen (11 men and 8 women) stable COPD patients from the pulmonary outpatient department of the University Hospital in Warsaw, Poland, with severe or very severe airway obstruction (FEV_1 < 50%), were recruited into the study in April–July 2016. All patients were treated with a long acting muscarinic antagonists and the majority received long acting beta-agonists (18/19; 95%) and inhaled corticosteroids (16/19; 84%). Four (21%) patients were on long term oxygen therapy. The baseline characteristics of the investigated group are presented in Table 1.

The major inclusion criteria were as follows: age ≥ 40 years, diagnosis of COPD in accordance with the GOLD recommendations (GOLD 2016), post-bronchodilator FEV_1 < 50% predicted and smoking history of at least 10 pack-years. The ability to remain in the supine position for at least 25 min (duration of a manual therapy session) also was mandatory. The patients remained on their usual inhaled therapy. The exclusion criteria comprised acute COPD exacerbation within 4 weeks before the study onset, exacerbation of any concomitant disease, acute rib or vertebral fracture, significant chest wall deformation (e.g., scoliosis) and inability to perform good quality, reproducible spirometry and body plethysmography.

The initial patient evaluation comprised medical history, spirometry with bronchial reversibility testing, COPD assessment test (CAT) (Jones et al. 2009), modified Medical Research Council dyspnea scale (mMRC).

2.3 Manual Therapy

Manual therapy (OMT or sham procedure) was applied in the supine position. Both the patient and the physician were blinded to the applied protocol while the therapist was responsible for implementing the appropriate manual therapy protocol in individual patients. The applied techniques for OMT and the sham protocol are presented in Table 2.

In all patients, the sessions were applied between 11.00 a.m. and 1 p.m. (i.e., at least 2 h after the intake of the morning dose of bronchodilators) on both observation days. The procedures were conducted by a trained therapist with at least 5-year experience in osteopathic manual therapy. In each patient, the exercise maneuvers were performed in the same order. The duration and number of repetitions of each technique depended on the improvement of fascial mobility and release muscle tension as judged by the therapist on palpation. Only then would the therapist apply the subsequent exercise as designed in the protocol. The duration of each session did not exceed 25 min.

Table 1 Clinical and functional characteristics of chronic obstructive pulmonary disease (COPD) patients at baseline, prior to any manual therapy intervention

Parameter	Value
Gender (M/F)	11/8
Age (years)	68 (63–72)
BMI (kg/m^2)	26.3 (23.1–27.95)
Ex-smokers/current smokers	17 (89.5)/ 2 (10.5)
Smoking history (pack-years)	40 (25–50.5)
Disease duration (years)	17 (10–24.5)
Baseline FEV$_1$ (% predicted)	43.4 (33.2–46.7)
Baseline FVC (% predicted)	87.8 (72.6–98.2)
Baseline RV (% predicted)	199.2 (173.9–231.3)
Baseline IC (% predicted)	96.4 (76.6–103.5)
Baseline FRC (% predicted)	167.5 (158.7–208.9)
Baseline TLC (% predicted)	139.7 (122.9–148.0)
Baseline RV/TLC (%)	58.2 (53.8–64.6)
Baseline IC/TLC (%)	30.8 (23.1–33.6)
Baseline FRC/TLC (%)	69.1 (66.4–76.8)
CAT score	18 (13–20)
mMRC	3 (3–4)
SpO$_2$ when breathing room air (%)	93 (92–97)

Results are presented as medians and interquartile ranges (IQR). *BMI* body mass index, *FEV$_1$* forced expiratory volume in one second, *FRC* functional residual capacity, *FVC* forced vital capacity, *IC* inspiratory capacity, *CAT* - COPD assessment test, *mMRC* modified Medical Research Council scale for dyspnea, *RV* residual volume, *TLC* total lung capacity, *SpO$_2$* peripheral capillary blood oxygen saturation measured by pulse oximetry

2.4 Pulmonary Function Testing and Evaluation of Dyspnea

Spirometry (Lung Test 1000, MES, Cracow, Poland) and body plethysmography (BodyBox 5500, Medi-Soft SA, Sorinnes, Belgium) were performed according to the recommendations of the European Respiratory Society and American Thoracic Society (Miller et al. 2005; Pellegrino et al. 2005; Wanger et al. 2005). Pulmonary function was assessed directly before and 10 min after each therapeutic session. In addition, before and after each session the perception of dyspnea was assessed on the visual analogue scale (VAS). A 10 cm horizontal line anchored between two marginal points which corresponded with the labels "I have no problems with breathing" (0 cm) and "I can hardly breathe at all" (10 cm) was used.

2.5 Statistical Analysis

The estimation of a sample size was based on data from the study by Noll et al. (2008). We assumed that when the last maneuver in the exercise of the thoracic lymphatic pump, consisting of the abrupt removal of the therapist's hands pressing on the pectoralis muscles to increase the volume of inspired air, is abandoned, RV would decrease, rather than increase as observed in the study mentioned above. Therefore, the assumed RV decrease following OMT would be at least 650 mL. An additional value of 100 mL (15%) was added to further emphasize the positive effect of OMT. To detect the difference of 750 mL with a power of 80% and a significance level of 5%, the sample size was estimated as 16 patients.

Data were presented as medians and IQR or percentages. Differences between continuous variables in the two groups were tested using the Mann–Whitney U test as data were not normally distributed. To determine significance of inter-differences, the sign test and signed rank test were applied. Differences were considered statistically significant at p < 0.05. The McNemar test was applied to determine the differences between the two therapeutic techniques. Statistical analyses were performed using Statistica v12.0

Table 2 Description of the osteopathic manual techniques and the sham intervention applied in chronic obstructive disease (COPD) patients

Osteopathic manual therapy protocol	Sham protocol
Suboccipital decompression	Shoulder joint mobilization by gliding techniques (glenohumeral anterior, posterior and inferior glide, circumduction)
Traction in the outward and cephalad direction with the fingers of the therapist placed on the occipital condyles at the base of the skull Aim: relaxation of the structures of the atlanto-occipital joint	Aim: improvement in shoulder flexion, extension, internal and external rotation
Deep cervical fascia release	Post-isometric relaxation of the shoulder rotators and the biceps brachii
Compression along the sternocleidomastoid, scalene, trapezius muscles from the base of the head in the caudal direction performed with the thumbs of the therapist Aim: improvement of the mobility of the upper ribs and relaxation of the anatomical regions associated with the vagus and phrenic nerve	Aim: increase in the mobility of the shoulder joint
Thoracic lymphatic pump	
Manual support of the patient's breathing rhythm by compression and decompression of the sternum effecting in the sterno-costal joints relaxation and intensification of lymph flow the thoracic duct. The impact on the thoracic duct is carried out by the rhythmic pressure differences while breathing and, in this case, supported with hands of the therapist Aim: reduction in the tension of mediastinal structures, improvement in lymphatic flow and thus reduction in breathing resistance	
Diaphragm "stretching"	
Repeated traction of the angles of the lower ribs in the outward (inspiratory) direction during expiration. The procedure was stopped when mobility of the lower ribs improved (arbitral decision of the therapist enhanced by palpation) Aim: improvement of the mobility of the lower part of the rib cage and reducing the adhesion between the diaphragm and the abdominal fascia	

Adapted from Noll et al. (2008)

software (StatSoft, Tulsa, OK) and Statgraphics Plus v.4.1. (Statpoint Technologies, Warrenton VA).

3 Results

3.1 Effect of Manual Therapy on Pulmonary Function and Dyspnea

All the included patients completed the study protocol. No adverse effects associated with the OMT and sham intervention were observed in any of the participating patients. With the exception of TLC, which was lower after the sham rehabilitation, neither pulmonary function nor dyspnea differed significantly before and after OMT and sham procedures (Table 3).

The differences in lung volumes before and after OMT and the sham protocol did not differ significantly. Of the 19 investigated patients, 7 (36.8%) demonstrated a decrease in RV (% predicted) exceeding the predetermined threshold of -6.1% after OMT. In comparison to patients who did not exceed this MID, the 'responders' had a greater perception of dyspnea at baseline – VAS score 7.0 (4.5–7.0) vs. 3.0 (0.0–5.0), respectively, p = 0.04. After the sham protocol, RV decreased by 6.1% in 9/19 (47.4%) patients and these patients were characterized by a significantly greater difference in the perception of

Table 3 Comparison of pulmonary function and perception of dyspnea before and after osteopathic manual therapy (OMT) and sham therapy

Variable	OMT-Pre n = 19	OMT-Post n = 19	p	Sham-Pre n = 19	Sham-Post n = 19	P
FEV_1 (L)	1.1 (0.8–1.4)	1.0 (0.7–1.3)	0.13	1.0 (0.7–1.3)	1.0 (0.8–1.3)	0.81
FEV_1% pred.	43.4 (34.5–46.7)	38.9 (33.6–45.8)	0.09	36.8 (33.1–46.4)	38.0 (33.6–47.4)	0.89
FVC (L)	2.9 (2.4–3.7)	3.2 (2.2–3.7)	0.72	3.0 (2.4–3.6)	2.9 (2.3–3.7)	0.30
FVC % pred.	87.7 (76.8–98.2)	87.5 (73.2–97.7)	0.55	85.9 (72.6–93.1)	85.5 (73.9–94.4)	0.40
FEV_1%FVC (%)	33.3 (29.4–43.1)	33.2 (30.0–43.3)	0.35	32.4 (29.2–43.1)	31.4 (28.5–43.7)	0.94
TLC (L)	7.5 (6.5–9.0)	7.5 (6.6–8.7)	0.28	**7.6 (6.9–9.0)**	**7.2 (6.5–8.8)**	**0.03**
TLC % of pred.	136.6 (122.9–147.8)	137.6 (121.0–144.9)	0.37	**139.9 (124.7–149.4)**	**138.3 (123.2–147.7)**	**0.03**
RV (L)	4.5 (3.8–4.9)	4.5 (3.8–4.8)	0.23	4.5 (4.2–5.2)	4.5 (4.1–5.1)	0.06
RV % pred.	198.8 (179.6–226.8)	196.5 (168.2–224.6)	0.35	207.9 (179.3–232.9)	198.8 (178.5–225.9)	0.08
RV/TLC (%)	57.5 (53.8–64.5)	57.0 (52.7–65.2)	0.28	59.6 (56.8–64.3)	58.6 (55.4–63.8)	0.16
Raw	3.1 (2.4–3.9)	2.8 (2.1–4.0)	0.44	2.8 (2.0–4.9)	2.8 (2.2–4.1)	0.98
Raw % of pred.	230 (178.8–290.1)	216.1 (167.5–286.9)	0.33	196.8 (166.3–363.5)	217.8 (175.6–272.3)	0.98
IC (L)	2.1 (1.92–2.7)	2.4 (1.8–2.7)	0.55	2.3 (1.9–2.7)	2.1 (1.7–2.7)	0.53
IC % of pred.	93.9 (77.6–107.4)	95.5 (84.0–107.1)	0.81	96.4 (81.4–106.9)	96.3 (84.8–107.7)	0.69
FRC (L)	5.6 (4.3–6.1)	5.7 (4.3–5.9)	0.16	5.5 (4.7–6.1)	5.3 (4.6–6.3)	0.09
FRC % of pred.	165.9 (157.9–204.8)	173.3 (148.2–197.2)	0.20	178.6 (162.2–196.9)	173.4 (154.3–190.8)	0.10
VAS	3.0 (0.5–6.0)	3.0 (1.0–4.5)	0.06	4.0 (1.0–5.5)	2.0 (1.0–4.0)	0.09

Results presented as median (IQR). FEV_1 forced expiratory volume in 1 s, *FVC* forced vital capacity, *TLC* total lung capacity, *RV* residual volume, *Raw* airway resistance, *IC* inspiratory capacity, *FRC* functional residual capacity, *VAS* visual analogue scale. Significant changes are depicted **in bold**

dyspnea before and after the session, compared to the 'non-responders' – Δ VAS −2.0 (−2.0 − {−1.0}) vs. 0.0 (−1.0 − {−1.0}), respectively, p = 0.02. Nonetheless, the overall pre- and post-intervention dyspnea did not differ significantly in the two groups.

No correlations between the clinical data, baseline pulmonary function values, and the median differences of lung volumes with their derivatives were found before and after OMT and sham therapy.

4 Discussion

This study did not show any significant influence of a single session of OMT on indices of hyperinflation and perception of dyspnea in patients with severe COPD. However, when we considered the established MID for RV in COPD patients (−6.1%), we found that patients in whom this MID was observed after the intervention were characterized by a significantly higher sense of dyspnea at baseline than those who did not achieve the MID. The adopted MID for RV was also reached by a group of patients who were subject to sham therapy involving the upper limbs, therefore a potential beneficial effect of manual therapy applied to the upper part of the body, not only to the chest wall *per se* may be expected.

The present findings are in contrast with the results of other authors. Two recent studies have shown that in patients with severe COPD, manual therapy applied in a single session may have a positive effect on indices of lung hyperinflation. Cruz-Montecinos et al. (2017) have demonstrated that manual therapy applied in one session resulted in a significant decrease in TLC, ERV and RV, which was accompanied by an increase in IC and SpO_2. Likewise, Yilmaz Yelvar et al. (2016) have noticed significant improvements in FEV_1, FVC, respiratory muscle strength, SpO_2 and a decrease in dyspnea perception after a single session comprising of soft tissue therapy and thoracic mobilization. These results support the use of manual therapy in the management of COPD and show that even a single intervention may have positive effects on breathing mechanics and symptom severity. The protocols used in the two mentioned studies were of longer duration than that in the present study (30 and 45 min vs. 25 min, respectively). There were also differences in the applied manual techniques and in gender distribution among the investigated patients, with a marked male predominance in both previous studies as opposed to equal gender distribution in the present study. Another difference was the younger age of patients in the previous studies and it is known that age affects the results of manual therapy (Engel and Vemulpad 2008). All these factors may have contributed to the difference in the outcomes. However, Noll et al. (2008) have found in a group of 18 patients with COPD that a 20-min OMT session not only failed to improve lung function, but even further increased hyperinflation and was also inferior to a sham procedure. Given the above, it seems that the duration of a single OMT session should be considered an important factor conditioning the effectiveness of this method of complimentary treatment in COPD patients.

The musculoskeletal responsiveness to manual therapy tends to decrease in the elderly due to age-related decrease in soft tissue elasticity (Bougie and Morgenthal 2001). Engel and Vemulpad (2008) have postulated that the elderly require more repetitions of manual therapy to achieve the effect attained with fewer applications in younger patients. However, the results of studies involving repeated manual therapy sessions in COPD patients are equivocal. Zanotti et al. (2012), who combined OMT with pulmonary rehabilitation and applied 45-min OMT sessions once a week for 4 weeks, have found that an increase in 6-min walking distance and an 11% decrease in RV, compared to baseline in patients with severe COPD. These changes were significantly more pronounced than those in patients in whom only pulmonary rehabilitation alone was applied. Engel et al. (2013) compared three groups of patients in whom three different interventions are used: soft tissue therapy alone, soft tissue therapy and spinal manipulation, and both combined with exercise applied during 8 consecutive sessions over a 4 week period. These authors showed that soft tissue therapy alone had a significantly worse outcome than the other two protocols. Another study by the same group of authors showed that manual therapy added to standard pulmonary rehabilitation resulted in a significantly higher FVC, compared to standard pulmonary rehabilitation alone and that the effect sustained after the termination of the intevention (Engel et al. 2016).

The present study has two major limitations. Firstly, we adopted an MID threshold which had been established for a different therapeutic method, namely bronchoscopic lung volume reduction (BLVR). However, this was imposed by the fact that from the static lung volumes, MID thresholds in patients with COPD have been proposed only for RV and RV/TLC and these values were calculated for patients undergoing BLVR (Hartman et al. 2012). There is a need for establishing the threshold for minimally (clinically) important difference for the most commonly applied functional indices (other than FEV_1), what would enable the evaluation of the effectiveness of therapeutic interventions in patients with COPD. Adopting the same MID threshold for two radically different treatment methods may be misleading and such a comparison ought to be interpreted with caution. The positive effect of a sham procedure may also be regarded a significant limitation of our study and raise doubts on the placebo effect of the sham techniques used in the study design. However, techniques involving the shoulder in the sham intervention were chosen to minimize the therapeutic effect on the chest wall on one side, and to focus on the anatomic region potentially associated with breathing to convince the patient that the session might have therapeutic implications on the other. A significant decrease in TLC observed in our patients after the sham protocol was not associated with a decrease in RV, FRC, and/or an increase in IC, so it seems an incidental finding. Further, the outcome measures were compared to the MID adopted from BVLR, with all the drawbacks involved as outlined above. Paradoxically, a positive response to the sham protocol involving the upper limbs in a subgroup of COPD patients may support the hypothesis that a beneficial effect of manual therapy applied to the upper part of the body, not only to the chest wall *per se*, may be expected.

Although the lack of benefit from a single OMT session in terms of improvement of pulmonary function in patients with severe COPD found in this study may seem somewhat discouraging, in our opinion, this complimentary method of COPD treatment need not be neglected. Manual therapy is not associated with significant adverse effects, activates the patient, necessitates a face-to-face interaction with the therapist, and may reduce the sense of social isolation, which is an important factor affecting quality of life of COPD patients (Meshe et al. 2017; Marx et al. 2016). Since there is evidence that repeated OMT sessions applied for a longer period of time (4 weeks) give benefit to hyperinflated COPD patients (Engel et al. 2013; Zanotti et al. 2012), the present study may be regarded as a negative pilot investigation that may be taken into consideration in the future design of intervention trials to modify entry criteria.

5   Conclusions

The results of this study did not confirm the immediate effect of a single session of osteopathic manual therapy protocol on hyperinflation and the sense of dyspnea in COPD patients with severe and very severe airway obstruction. In the light of these observations and the findings of earlier studies, further research on the influence of manual therapy as a complimentary method for COPD treatment seems warranted. Further exploration of the issue ought to focus on targeted subgroups of patients, optimal techniques and their duration to be applied, and the search for optimal objective outcome measures.

Conflicts of Interest The authors declare no conflicts of interest in relation to this article.

Ethical Approval All procedures performed in studies involving human participants were in accordance with the ethical standards of the institutional and/or national research committee and with the 1964 Helsinki declaration and its later amendments or comparable ethical standards. The project was approved by the institutional Bioethics Committee of the Medical University of Warsaw in Poland and it was registered at ClinicalTrial.gov (NCT02755363).

Informed Consent Written informed consent was obtained from all individual participants included in the study.

References

Bolton CE, Bevan–Smith EF, Blakey JD et al (2013) British Thoracic Society Pulmonary Rehabilitation Guideline Development Group; British Thoracic Society Standards of Care Committee. British Thoracic Society guideline on pulmonary rehabilitation in adults. Thorax 68:ii1–i30

Bougie JD, Morgenthal AP (2001) The aging body: conservative management of common neuromusculoskeletal conditions. McGraw Hill Medical, New York

Celli BR (2017) Pulmonary rehabilitation. Up–to–Date. https://www.uptodate.com/contents/pulmonary–rehabilitation. Accessed 23 on Feb 2018

Cicchitti L, Martelli M, Cerritelli F (2015) Chronic inflammatory disease and osteopathy: a systematic review. PLoS One 10(3):e0121327

Cruz–Montecinos C, Godoy–Olave D, Contreras–Briceño FA, Gutiérrez P, Torres–Castro R, Miret–Venegas L,

Engel RM (2017) The immediate effect of soft tissue manual therapy intervention on lung function in severe chronic obstructive pulmonary disease. Int J Chron Obstruct Pulmon Dis 12:691–696

Engel RM, Vemulpad SR (2008) Immediate effects of osteopathic manipulative treatment in elderly patients with chronic obstructive pulmonary disease. J Am Osteopath Assoc 108(10):541–542

Engel RM, Vemulpad SR, Beath K (2013) Short–term effects of a course of manual therapy and exercise in people with moderate chronic obstructive pulmonary disease: a preliminary clinical trial. J Manip Physiol Ther 36(8):490–496

Engel RM, Gonski P, Beath K, Vemulpad S (2016) Medium term effects of including manual therapy in a pulmonary rehabilitation program for chronic obstructive pulmonary disease (COPD): a randomized controlled pilot trial. J Man Manip Ther 24(2):80–89

GOLD (2016) Global initiative for chronic obstructive lung disease. Global strategy for the diagnosis, management and prevention of COPD. https://www.goldcopd.org/global–strategy–diagnosis–management–prevention–copd–2016. Accessed on 30 Jan 2018

GOLD (2019) Global initiative for chronic obstructive lung disease. Global strategy for the diagnosis, management and prevention of COPD. http://www.goldcopd.org. Accessed on 18 Mar 2019

Hartman JE, Ten Hacken NH, Klooster K, Boezen HM, de Greef MH, Slebos DJ (2012) The minimal important difference for residual volume in patients with severe emphysema. Eur Respir J 40:1137–1141

Heneghan NR, Adab P, Balanos GM, Jordan RE (2012) Manual therapy for chronic obstructive airways disease: a systematic review of current evidence. Man Ther 17:507–518

Hogg L, Garrod R, Thornton H, McDonnell L, Bellas H, White P (2012) Effectiveness, attendance, and completion of an integrated, system–wide pulmonary rehabilitation service for COPD: prospective observational study. COPD 9:546–554

Howell RK, Allen TW, Kappler RE (1975) The influence of osteopathic manipulative therapy in the management of patients with chronic obstructive pulmonary disease. J Am Osteopath Assoc 74:757–760

Jones PW, Harding G, Berry P, Wiklund I, Chen WH, Kline Leidy N (2009) Development and first validation of the COPD assessment test. Eur Respir J 34:648–654

Langer D, Ciavaglia CE, Neder JA, Webb KA, O'Donnell DE (2014) Lung hyperinflation in chronic obstructive pulmonary disease: mechanisms, clinical implications and treatment. Expert Rev Respir Med 8:731–749

Marx G, Nasse M, Stanze H, Boakye SO, Nauck F, Schneider F (2016) Meaning of living with severe chronic obstructive lung disease: a qualitative study. BMJ Open 6(12):e011555

Meshe OF, Claydon LS, Bungay H, Andrew S (2017) The relationship between physical activity and health status in patients with chronic obstructive pulmonary disease following pulmonary rehabilitation. Disabil Rehabil 39:746–756

Miller MR, Hankinson J, Brusasco V et al (2005) ATS/ERS task force. Standardisation of spirometry. Eur Respir J 26:319–338

Morais N, Cruz J, Marques A (2016) Posture and mobility of the upper body quadrant and pulmonary function in COPD: an exploratory study. Braz J Phys Ther 20:345–354

Noll DR, Degenhardt BF, Johnson JC, Burt SA (2008) Immediate effects of osteopathic manipulative treatment in elderly patients with chronic obstructive pulmonary disease. J Am Osteopath Assoc 108:251–257

Pellegrino R, Viegi G, Brusasco V, Crapo RO, Burgos F, Casaburi R, Coates A, van der Grinten CP, Gustafsson P, Hankinson J, Jensen R, Johnson DC, MacIntyre N, McKay R, Miller MR, Navajas D, Pedersen OF, Wanger J (2005) Interpretative strategies for lung function tests. Eur Respir J 26:948–968

Steel A, Sundberg T, Reid R, Ward L, Bishop FL, Leach M, Cramer H, Wardle J, Adams J (2017) Osteopathic manipulative treatment: a systematic review and critical appraisal of comparative effectiveness and health economics research. Musculoskelet Sci Pract 27:165–175

Wanger J, Clausen JL, Coates A, Pedersen OF, Brusasco V, Burgos F, Casaburi R, Crapo R, Enright P, van der Grinten CP, Gustafsson P, Hankinson J, Jensen R, Johnson D, Macintyre N, McKay R, Miller MR, Navajas D, Pellegrino R, Viegi G (2005) ATS/ERS task force. Standardisation of the measurement of lung volumes. Eur Respir J 26:511–522

Yilmaz Yelvar GD, Cirak Y, Demir Y, Dalikilinc M, Bozkurt B (2016) Immediate effect of manual therapy on respiratory functions and inspiratory muscle strength in patients with COPD. Int J Chron Obstruct Pulmon Dis 11:1353–1357

Zanotti E, Berardinelli P, Bizzarri C, Civardi A, Manstretta A, Rossetti S, Fracchia C (2012) Osteopathic manipulative treatment effectiveness in severe chronic obstructive pulmonary disease: a pilot study. Complement Ther Med 20:16–22

Advs Exp. Medicine, Biology - Neuroscience and Respiration (2019) 44: 27–35
https://doi.org/10.1007/5584_2019_433
© Springer Nature Switzerland AG 2019
Published online: 27 September 2019

Oxidative Stress Markers and Severity of Obstructive Sleep Apnea

S. Cofta, H. M. Winiarska, A. Płóciniczak, L. Bielawska, A. Brożek, T. Piorunek, T. M. Kostrzewska, and E. Wysocka

Abstract

Oxidative stress underlies both obstructive sleep apnea (OSA) and atherosclerosis. The aim of the study was to assess the markers of oxidative stress in plasma at different stages of OSA in non-smoking obese Caucasian males aged 41–60, with normal oral glucose tolerance test. All patients were subjected to clinical and polysomnographic examinations. The stage of OSA severity was set according to the following criteria of the apnea-hypopnea index (AHI): AHI < 5/h – no disease (OSA-0; n = 26), AHI 5–15/h – mild disease (OSA-1; n = 26), AHI 16–30/h – moderate disease (OSA-2: n = 27), and AHI > 30/h obstructive episodes per hour – severe disease (OSA-3; n = 27). Plasma total antioxidant status (TAS) and thiobarbituric acid-reacting substances (TBARS), reflecting the level of lipid peroxides, were determined spectrophotometrically. We found that TAS decreased and TBARS increased significantly from OSA-0 to OSA-3. We conclude that the oxidative stress markers are conducive to setting the severity of OSA in normoglycemic patients.

Keywords

Antioxidant activity · Apnea/hypopnea · Atherosclerosis · Obstructive sleep apnea · Oxidative stress · Polysomnography

1 Introduction

Obstructive sleep apnea (OSA) is a pathology of breathing during sleep and it is characterized by the recurrent episodes of apneas or hypopneas, caused by upper airway obstruction, usually with coexisting functional or anatomical disorders. The OSA prevalence in adult population is about 2–4%. In obese people the incidence reaches up to 40–70% (Framnes and Arble 2018). The most important risk factors are: middle age, male gender, central body fat deposition, anatomical upper airway disorders, overuse of alcohol and some medicaments, smoking, and physical inactivity (Somers et al. 2008). Symptoms, such as increased daytime sleepiness, falling asleep during routine activities, chronic fatigue, and social withdrawal, make OSA an important personal and social problem.

OSA can be diagnosed and graded by polysomnography. Current classification of OSA is based on apnea/hypopnea index (AHI) to

S. Cofta (✉), H. M. Winiarska, T. Piorunek, and T. M. Kostrzewska
Department of Respiratory Medicine, Allergology and Pulmonary Oncology, Poznań University of Medical Sciences, Poznań, Poland
e-mail: scofta@ump.edu.pl

A. Płóciniczak, L. Bielawska, and E. Wysocka
Department of Laboratory Diagnostics, Poznań University of Medical Sciences, Poznań, Poland

A. Brożek
Department of Clinical Biochemistry and Laboratory Medicine, Poznań University of Medical Sciences, Poznań, Poland

recognize mild, moderate, and severe stage of the disease. Consequences of untreated OSA include ischemic heart disease, stroke, arrhythmia, chronic heart failure and are related to increased cardiovascular mortality (McNicholas et al. 2007). Moreover, these diseases could be also discussed as atherosclerotic and low-grade chronic inflammatory disorders.

Cardiovascular complications are the main cause of death in patients suffering from OSA. The difference in all-cause mortality between OSA and non-OSA male patients is seen especially in individuals aged under 50 (Lavie et al. 1995; He et al. 1988). It is still necessary to recognize the mechanisms linking OSA with the development of atherosclerosis; with oxidative stress potentially being one of them (Lavie 2009; Cofta et al. 2008). Repeated apnea/hypopnea episodes disorder the sleep architecture, cause intra-thoracic pressure fluctuations, cyclical alterations of arterial oxygen saturation, sympathetic activation, and increase oxidative stress, all of which leads to tissue damage and organ failures and to the acceleration of atherosclerosis (Bonsignore and Zito 2008). Therefore, the aim of this study was to assess the plasma content of oxidative stress markers, such as total antioxidant status (TAS) and thiobarbituric acid-reacting substances (TBARS), in different stages of OSA in an attempt to get closer insight into the association between OSA severity and increased risk of atherosclerosis.

2   Methods

One hundred and six men, aged 34–64, with a clinical suspicion of OSA, referred for testing in the Sleep Laboratory of the Department of Pulmonology, Allergology and Respiratory Oncology of Poznań University of Medical Sciences in Poland were enrolled into this study. Exclusion criteria were the following: chronic and acute heart diseases, stroke, respiratory failure, chronic kidney disease, chronic liver disease, current smoking, cancer, diabetes mellitus, and elevated

fasting glycemia or impaired glucose tolerance.

Patients underwent a physical examination and had the day time sleepiness estimated on the Epworth Sleepiness Scale (ESS). Then, they underwent a full-night monitoring by the polysomnographic system EMBLA S4000-Remlogic equipped with Somnologica Studio v5.0 software (Embla Systems Inc., Thornton, CO). Computer recordings were manually reevaluated for verification. Apnea was defined as a cessation of airflow lasting more than 10 s and hypopnea as $\geq 30\%$ reduction of airflow with $\geq 4\%$ desaturation, lasting for more than 10 s.

The study population was stratified by AHI, based on the American Academy of Sleep Medicine Clinical Practice Guideline (Kapur et al. 2017). There were 26 patients with AHI < 5 episodes/h of sleep, who were judged OSA negative (OSA-0); another 26 patients with mild OSA of AHI 5–15/h (OSA-1); 27 with moderate OSA of AHI 16–30/h (OSA-2); and another 27 with severe OSA of AHI > 30/h (OSA-3).

After clinical examination a set of fasting blood tests were performed. A complete blood count was measured using the Advia 2120i analyzer (Siemens Healthcare Diagnostics; Deerfield, IL). Oral glucose tolerance test (OGTT), including fasting (G-0′) and 120-min glycemia (G-120′), was carried out if there were no contraindications. Total cholesterol (T-C), HDL-cholesterol (HDL-C), LDL-cholesterol (LDL-C), triglycerides (TAG), high-sensitivity C-reactive protein (hsCRP), as well as glucose concentrations were measured using Dimension Xpand Plus Systems (Siemens Healthcare Diagnostics; Deerfield, IL). The patients with the white blood cell count (WBC) $< 9.00 \times 10^9$/l, hsCRP < 8.00 mg/l and with normal glucose tolerance were enrolled.

Total antioxidant status (TAS) was assessed by the colorimetric method using Randox Reagent Kit (Randox Laboratories; Crumlin, UK) and Statfax^{tm} 1904 Plus spectrophotometer (Awareness Technology; Pal City, FL). The intra- and inter-assay coefficients of variation (CV) were 1.5% and 3.8%, respectively. The thiobarbituric acid-reacting substances (TBARS)

– interpreted as products of lipid peroxidation in plasma (Youseff et al. 2014) – were measured by a colorimetric method using Sigma-Aldrich (St. Louis, MO) reagents and a Specord M40 spectrophotometer (Zeiss-Analytik Jena AG; Jena, Germany). The intra- and inter-assay CV were 1.8% and 3.7%, respectively.

Data were expressed as medians with interquartile ranges (25–75%). The Shapiro-Wilk test was used to check the normality of data distribution in the groups studied and consequently non-parametric tests were used in further analysis. Inter-group differences were compared with the analysis of variance by the Kruskal-Wallis test and *post hoc* multiple comparisons by the Dunn test. Correlations were assessed using the critical values adopted by Spearman. Multiple regression analysis was conducted for relevant variables. A p-value $p < 0.05$ was taken as an indicator of statistical significance of differences. The analysis was performed using a commercial statistical package of Statistica v10.0 (StatSoft; Tulsa, OK).

3 Results

3.1 Patients Characteristics

Clinical and laboratory characteristics of the OSA subgroups and relevant comparisons among them are shown in Table 1. Results of multiple comparison are indicated in the text below, whenever appropriate. By definition, the groups differed in the AHI and ESS values. The patients presented a comparable BMI (28.6–31.6 kg/m^2). However, a tendency for increased neck and waist circumference became significant between OSA-1 – OSA-2 and OSA-2 – OSA-3 ($p = 0.034$ and $p = 0.003$, respectively). There were significant increases in blood glucose and hsCRP, along with decreases in HDL-C, in OSA-3 compared to OSA-0 (G-0′ $p = 0.014$, hsCRP $p = 0.020$, HDL-C; $p = 0.006$) and to OSA-1 (hsCRP $p = 0.014$, HDL-C; $p = 0.021$).

Table 1 Clinical and laboratory characteristics of patients in the successive subgroups of obstructive sleep apnea (OSA) severity and the overall significance of intra-group differences

	OSA-0 ($n = 26$)	OSA-1 ($n = 26$)	OSA-2 ($n = 27$)	OSA-3 ($n = 27$)	Significance
AHI (events/h)	2.2 (1.5–3.4)	7.8 (7.1–11.1)	21.0 (17.0–26.0)	49.6 (36.9–65.4)	By definition
Age (years)	53.5 (46.0–59.0)	53.0 (43.0–58.0)	54.0 (51.0–58.0)	54.0 (41.0–60.0)	ns
ESS (points)	4.0 (2.0–6.0)	4.5 (4.0–6.0)	9.0 (5.0–11.0)	10.0 (6.0–12.0)	$p < 0.0001$
BMI (kg/m^2)	28.6 (26.3–32.5)	28.8 (27.0–32.3)	30.1 (27.3–32.2)	31.6 (28.1–35.9)	ns
Waist (cm)	100 (94–108)	100 (92–110)	104 (96–110)	108 (100–118)	$p = 0.022$
Neck (cm)	42 (41–43)	41 (40–43)	42 (41–44)	44 (42–46)	$p = 0.005$
SBP (mmHg)	130 (120–140)	130 (120–140)	130 (130–140)	130 (130–135)	ns
DBP (mmHg)	80 (80–90)	80 (80–90)	80 (80–90)	80 (80–85)	ns
G-0′ (mmol/l)	5.1 (4.9–5.3)	5.1 (4.9–5.5)	5.3 (5.1–5.5)	5.4 (5.2–5.6)	$p = 0.017$
G-120′ (mmol/l)	6.2 (5.7–6.9)	5.9 (5.3–6.3)	5.9 (5.5–7.1)	6.5 (5.7–6.9)	ns
T-C (mmol/l)	5.0 (4.2–5.60)	5.3 (4.5–6.1)	5.3 (4.7–5.9)	5.2 (4.5–6.1)	ns
TAG (mmol/l)	1.7 (1.1–1.9)	1.3 (1.0–1.8)	1.7 (0.8–2.2)	1.5 (1.1–2.3)	ns
HDL-C (mmol/l)	1.3 (1.1–1.4)	1.2 (1.0–1.5)	1.2 (1.0–1.3)	1.0 (0.9–1.1)	$p = 0.004$
LDL-C (mmol/l)	2.9 (2.4–3.3)	3.4 (2.5–4.0)	3.3 (2.8–3.9)	3.4 (2.3–4.0)	ns
Non-HDL-C (mmol/l)	3.6 (2.9–4.6)	4.1 (3.1–4.7)	3.8 (3.5–5.0)	4.1 (3.3–5.0)	ns
WBC (10^9/l)	6.3 (5.6–7.3)	6.2 (5.6–6.9)	6.8 (5.3–7.3)	7.2 (6.1–7.8)	ns
hsCRP (mg/l)	2.0 (1.1–2.7)	1.9 (0.6–3.1)	2.8 (1.5–4.8)	3.4 (1.7–5.0)	$p = 0.003$

Data are presented as medians (interquartile range); *ns* non-significant, *AHI* apnea/hypopnea index, *BMI* body mass index, *ESS* Epworth's sleepiness scale, *SBP* systolic blood pressure, *DBP* diastolic blood pressure, *G-0′* fasting glycemia, *G-120′* 120-minute glycemia during oral glucose tolerance test, *T-C* total cholesterol, *TAG* triglycerides, *HDL-C* high density lipoprotein-cholesterol, *LDL-C* low density lipoprotein-cholesterol, *non-HDL-C* non-high density lipoprotein-cholesterol, *WBC* white blood cells, *hsCRP* high sensitivity C-reactive protein; p-value for Kruskal-Wallis test

3.2 Oxidative Stress Markers

TAS decreased from OSA-0 to OSA-3 (Fig. 1), with the following *post-hoc* inter-group differences: OSA-0 vs. OSA-1 (p = 0.008), OSA-0 vs. OSA-2 (p < 0.0001), OSA-0 vs. OSA-3 (p < 0.0001), OSA-1 vs. OSA-3 (p = 0.023). On the other side, TBARS increased in successive subgroups of OSA severity (Fig. 2): OSA-0 vs. OSA-2, OSA-0 vs. OSA-3, OSA-1 vs. OSA-2, OSA-1 vs. OSA-3 (p < 0.0001 for all). Accordingly, TAS/TBARS ratio significantly decreased with the severity of OSA (Fig. 3).

Taking into account all of the OSA patients (n = 80), we noticed a weak-to-moderate adverse relationship between the TAS content, on the one side, and neck and waist circumference, ESS, and AHI, on the other side. In contradistinction, there

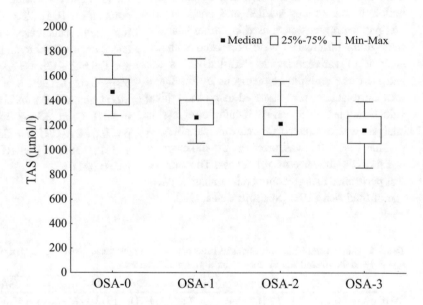

Fig. 1 Total antioxidant status (TAS) in the successive subgroups of obstructive sleep apnea (OSA) severity; *OSA-0,* disease negative patients, *OSA-1,* AHI 5–15/h, *OSA-2,* AHI 16–30/h, *OSA-3,* AHI > 30/h; *Min,* minimum; *Max,* maximum; *25%–75%,* interquartile range; *p < 0.0001,* Kruskal-Wallis test

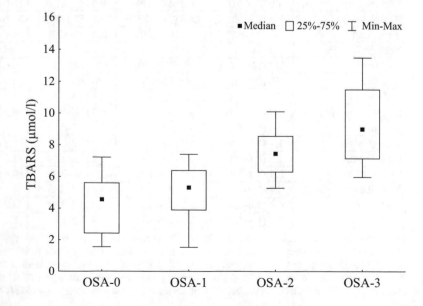

Fig. 2 Thiobarbituric acid-reacting substances (TBARS) in the successive subgroups of obstructive sleep apnea (OSA) severity; *OSA-0,* disease negative patients, *OSA-1,* AHI 5–15/h, *OSA-2,* AHI 16–30/h, *OSA-3,* AHI > 30/h; *Min,* minimum; *Max,* maximum; *25%–75%,* interquartile range; *p < 0.0001,* Kruskal-Wallis test

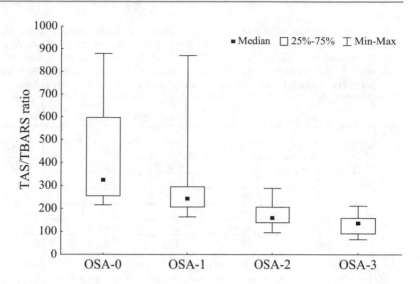

Fig. 3 Total antioxidant status (TAS)/thiobarbituric acid-reacting substances (TBARS) ratio in the successive subgroups of obstructive sleep apnea (OSA) severity; *OSA-0,* disease negative patients, *OSA-1,* AHI 5–15/h, *OSA-2,* AHI 16–30/h, *OSA-3,* AHI > 30/h; *Min,* minimum; *Max,* maximum; *25%–75%,* interquartile range; *p < 0.0001,* Kruskal-Wallis test

was a moderate-to-strong positive relationship between the TBARS content and ESS, BMI, neck and waist circumference, and AHI. In the OSA-0 group (non-OSA patients), systolic blood pressure was the only factor adversely related to TAS, while neck and waist circumference, BMI, and hsCRP were positively related to TBARS. In OSA-1 group, more factors, including the anthropometric features and SBP and DBP, positively related to TBARS, while AHI and ESS adversely related to TAS. Relations between TAS and TBARS and the afore-mentioned factors were rather irregular in more advanced stages of OSA.

The TAS/TBARS ratio could be proposed as an expression of a final antioxidant/oxidant balance in blood plasma. This balance was tilted toward TBARS with increasing severity of OSA and TBARS determined the ratio to a large extent in each OSA group investigated. An unfavorable relationship was noticed between oxidative stress markers and the severity of sleep apnea, expressed by AHI and clinical signs of the disease, which included a decrease in TAS and an increase in TBARS (Table 2).

Multiple regression analysis was used to assess to what extent the OSA-related variables could independently influence the oxidative stress markers. Details are presented in Table 3. The ESS contributed to about 25% of plasma TAS variability in OSA-1 and neck circumference to

37% of TAS variability in OSA-2. Taking into account all OSA patients (OSA-1, 2, 3), increased AHI contributed to 26% of a decrease in TAS. The BMI and waist circumference contributed to almost 81% of TBARS variability in non-OSA patients, and BMI still affected 51% of TBARS variability OSA-1 group. Likewise, AHI strongly associated with TBARS, contributing to about 85% of its variability in OSA-3 patients. Finally, BMI affected, in an adverse manner, 63% of TAS/TBARS ratio variability in non-OSA persons. In OSA-2 and OSA-3 patients, it was the AHI that contributed to TAS/TBARS variability the most; by 40% and 70%, respectively. To sum up, in normoglycemic non-OSA and mild OSA patients, excess body mass was a key factor affecting TBARS, while in severe OSA, the AHI contributed to TBARS and TAS/TBARS ratio changes, independently from other factors.

4 Discussion

Oxidative stress plays a role in the development of a spate of disorders, notably vasculopathies, including ischemic heart disease, hypertension, arrhythmias and chronic heart failure, all of which result from the activation of inflammatory pathways, endothelial dysfunction, and

Table 2 Relationships of the oxidative stress markers (total antioxidant status – TAS, thiobarbituric acid-reacting substances – TBARS, and TAS/TBARS ratio) with clinical and biochemical patient characteristics in the successive subgroups of obstructive sleep apnea (OSA) severity: *OSA-0,* disease negative patients, *OSA-1,* AHI 5–15/h, *OSA-2,* AHI 16–30, *OSA-3,* AHI > 30/h, described by Spearman's correlation coefficient and included in the table if significant (p < 0.05) at least in one of the study subgroups

	OSA-0 (n = 26)	OSA-1 (n = 26)	OSA-2 (n = 27)	OSA-3 (n = 27)	OSA-1, 2, 3 (n = 80)	All patients (n = 106)
TAS vs.						
AHI	ns	−0.398	ns	ns	−0.416	−0.632
ESS	ns	−0.532	−0.397	ns	−0.316	−0.401
Waist	ns	ns	ns	ns	−0.231	−0.212
Neck	ns	ns	−0.574	ns	−0.373	−0.290
SBP	−0.403	ns	ns	ns	ns	ns
TBARS	ns	ns	ns	ns	−0.383	−0.514
TAS/TBARS	ns	ns	0.787	0.656	0.597	0.699
TBARS vs.						
AHI	ns	ns	0.537	0.949	0.779	0.803
ESS	ns	ns	ns	ns	0.359	0.460
BMI	0.888	0.721	ns	ns	0.371	0.461
Waist	0.754	0.476	ns	ns	0.393	0.458
Neck	0.436	0.403	ns	0.447	0.497	0.436
SBP	ns	0.421	ns	ns	ns	0.244
DBP	ns	0.514	ns	ns	ns	ns
HDL-C	ns	ns	ns	ns	ns	−0.297
hsCRP	0.603	ns	ns	ns	ns	0.298
TAS/TBARS	−0.942	−0.942	−0.845	−0.915	−0.951	−0.961
TAS/TBARS vs.						
AHI	ns	ns	−0.538	−0.838	−0.804	−0.848
ESS	ns	ns	ns	ns	−0.383	−0.476
BMI	−0.832	−0.621	−0.411	ns	−0.367	−0.427
Waist	−0.659	ns	−0.443	ns	−0.422	−0.445
Neck	−0.448	ns	−0.410	−0.503	−0.542	−0.445
SBP	ns	ns	ns	ns	ns	−0.266
DBP	ns	−0.454	ns	ns	ns	ns
HDL-C	ns	ns	ns	−0.405	ns	0.290
hsCRP	−0.679	ns	ns	ns	−0.226	−0.310

AHI apnea/hypopnea index, *ESS* Epworth sleepiness scale, *Waist* waist circumference, *Neck* neck circumference, *BMI* body mass index, *SBP* systolic blood pressure, *DBP* diastolic blood pressure, *HDL-C* high density lipoprotein-cholesterol, *hsCRP* high sensitivity C-reactive protein, *ns* nonsignificant

consequently atherosclerosis (Griendling and FitzGerald 2003). Oxidative stress and systemic inflammation also are recognized as the fundamental mechanisms underlying cardiovascular morbidity and in OSA. According to Lavie and Lavie (2009) at least 50% of OSA patients are hypertensive and 10–15% have other cardiovascular disorders. The incidence of both fatal and non-fatal cardiovascular events in OSA is much greater than that in other patient populations (Marin et al. 2005, Robinson et al. 2004). Cardiovascular sequelae of OSA are underlain by increased sympathetic tone, hypertension, pro-inflammatory cytokine production, and metabolic dysregulation (Somers et al. 2008). Lavie (2009) has reported that OSA patients, even without any cardiovascular symptoms exhibit subclinical signs of arteriosclerosis such as increased

Table 3 Multivariable regression analysis involving each oxidative stress marker (total antioxidant status – TAS, thiobarbituric acid-reacting substances – TBARS, and TAS/TBARS ratio) as an outcome variable and clinical and biochemical patient characteristics as predictor variables in the successive subgroups of obstructive sleep apnea (OSA) severity: OSA-0, disease negative patients, *OSA-1*, AHI 5–15/h, *OSA-2*, AHI 16–30/h, *OSA-3*, AHI > 30/h. Variables described with the β coefficient significantly correlated, independently from the other variables, with a given stress marker

Outcome variable	OSA-0 ($n = 26$)	OSA-1 ($n = 26$)	OSA-2 ($n = 27$)	OSA-3 ($n = 27$)	OSA-1, 2, 3 ($n = 80$)
TAS	n/a	p = 0.013 for: AHI ESS (β = −0.490; $R^2 = 0.255$)	p = 0.004 for: ESS Neck (β = −0.530; $R^2 = 0.374$)	n/a	p = 0.0001 for: AHI (β = −0.330; $R^2 = 0.259$) ESS Waist Neck
TBARS	p < 0.0001 for: BMI (β = 0.571; $R^2 = 0.809$) Waist (β = 0.421; $R^2 = 0.809$) Neck hsCRP	p = 0.001 for: BMI (β = 0.572; $R^2 = 0.507$) Waist Neck SBP DBP	n/a	p < 0.0001 for: AHI (β = 0.981; $R^2 = 0.845$), Neck	p < 0.0001 for: BMI Waist Neck ESS (β = 0.258; $R^2 = 0.298$)
TAS/ TBARS	p − 0.0002 for: BMI (β = −0.450; $R^2 = 0.629$) Waist Neck hsCRP	ns BMI DBP	p = 0.019 for: AHI (β = −0.450; $R^2 = 0.400$) BMI Waist Neck	p < 0.0001 for: AHI (β = −0.740; $R^2 = 0.702$) Neck HDL-C	p < 0.0001 for: AHI (β = −0.430; $R^2 = 0.276$) ESS BMI Waist Neck hsCRP

β, standardized beta coefficient indicating how strongly a predictor variable independently influences the outcome variable, *n/a,* non-applicable, i.e., no variable was related to a given stress marker, *AHI* apnea/hypopnea index, *ESS* Epworth sleepiness scale, *Waist* waist circumference, *Neck* neck circumference, *BMI* body mass index, *SBP* systolic blood pressure, *DBP* diastolic blood pressure, *HDL-C* high density lipoprotein-cholesterol, *hsCRP* high sensitivity C-reactive protein, *ns* nonsignificant

intima-media thickness and arterial plaque formation. McNicholas et al. (2007) have reported that OSA is a significant and independent risk factor of all cerebrovascular and cardiovascular events and that CPAP therapy reduces cardiovascular morbidity and mortality to the level present in the general population.

Dyugovskaya et al. (2002) have reported an increased production of reactive oxygen species (ROS) in leukocytes of patients with OSA. Hypoxia-reoxygenation cycles generate overabundance of ROS, shifting the redox balance toward the oxidative side, which is conducive to inflammation. Another consequence of increased ROS production is enhanced oxidation of lipids. In the present study we found an increase in lipid peroxidation products with increasing severity of OSA, as evidenced by a higher level of TBARS, which is

in line with previous studies on the subject (Lavie and Lavie 2009; Barceló et al. 2000). Hopps et al. (2014) have reported that the incidence of cardiovascular episodes is associated with the severity stage of OSA, which we found in this study, is paralleled by a steady decrease in the TAS/TBARS index, i.e., a shift to the oxidative state. TBARS, a denominator in this ratio and an accepted, albeit indirect, measure of the intensity of oxidative stress leading to atherogenic events, was a predominant determinant of the decrease. This finding is in line with those studies that show mutual increases in TBARS content and AHI severity in untreated OSA patients (Hopps et al. 2014; Barceló et al. 2000) or decreases in response to CPAP treatment (Celec 2012; Barceló et al. 2006). The issue of TBARS changes in OSA is somehow contentious as there are studies contradicting any appreciable changes of this indicator of lipid peroxidation in OSA patients (Ntalapascha et al. 2012; Svatikova et al. 2005). The divergent results spurred the assessment of oxidative stress in OSA using alternative methods. Passali et al. (2015) have investigated the content of non-protein bound iron and isoprostanes and show that both are increased in OSA versus non-OSA patients. Sales et al. (2013) have reported that OSA patients have lower levels of witamin E and superoxide dismutase, but a higher level of homocysteine. Baysal et al. (2012) have reported that the level of ceruloplasmin and lipid peroxides also are increased in OSA patients. Further, other studies have pointed to a probable downturn in the antioxidative capability, expressed in practice by total antioxidant status in blood plasma (Erel 2004), present at the time of increased prooxidant propensity in OSA patients (Cofta et al. 2008). In the present study we confirmed the presence of a significant decrease in TAS with increasing severity of OSA, expressed by increased AHI. We also showed that AHI contributed to TAS/TBARS ratio variability in 40% and 70% in moderate and severe OSA, respectively.

In synopsis, decreasing TAS and increasing TBARS plasma content correlated with increases in OSA intensity and pointed to the atherogenic risk profile. That leads to a conclusion that the oxidative stress markers might play a favorable role in the categorization of metabolic characteristics of OSA, especially in relation to OSA severity. These markers could complement the assessment of a cardiovascular risk related to OSA pathology.

Conflict of Interest The authors declare no conflicts of interests in relation to this article.

Ethical Approval All procedures performed in studies involving human participants were in accordance with the ethical standards of the institutional and/or national research committee and with the 1964 Helsinki declaration and its later amendments or comparable ethical standards. The study was approved by the Ethics Committee of Poznań University of Medical Sciences in Poland.

Informed Consent Written informed consent was obtained from all individual participants included in the study.

References

Barceló A, Barbe F, de la Pena M, Vila M, Perez G, Pierola J, Duran J, Augusti AG (2000) Abnormal lipid peroxidation in patients with sleep apnoea. Eur Respir J 16:644–647

Barceló A, Barbe F, de la Pena M, Vila M, Perez G, Pierola J, Duran J, Augusti AG (2006) Antioxidant status in patients with sleep apnoea and impact of continuous positive airway pressure treatment. Eur Respir J 27:756–760

Baysal E, Taysi S, Aksoy N, Uyar M, Celenk F, Karatas ZA, Tarakcioglu M, Bilinç H, Mumbuç S, Kanlikama M (2012) Serum paraoxonase, arylesterase activity and oxidative status in patients with obstructive sleep apnea syndrome (OSAS). Eur Rev Med Pharmacol Sci 16:770–774

Bonsignore MR, Zito A (2008) Metabolic effects of the obstructive sleep apnea syndrome and cardiovascular risk. Arch Physiol Biochem 114:255–260

Celec P, Hodosy J, Behuliak M, Pálffy R, Gardlík R, Halčák L, Mucska I (2012) Oxidative and carbonyl stress in patients with obstructive sleep apnea treated with continuous positive airway pressure. Sleep Breath 16:393–398

Cofta S, Wysocka E, Piorunek T, Rzymkowska M, Batura–Gabryel H, Torliński L (2008) Oxidative stress markers in the blood of persons with different stages of obstructive sleep apnea syndrome. J Physiol Pharmacol 59(Suppl6):183–190

Dyugovskaya L, Lavie P, Lavie L (2002) Increased adhesion molecules expression and production of reactive oxygen species in leukocytes of sleep apnea patients. Am J Respir Crit Care Med 165:934–939

Erel O (2004) A novel automated direct measurement method for total antioxidant capacity using a new generation, more stable ABTS radical cation. Clin Biochem 37:277–285

Framnes SV, Arble DM (2018) The bidirectional relationship between obstructive sleep apnoea and metabolic disease. Front Endocrinol (Lausanne) 9:440

Griendling KK, FitzGerald GA (2003) Oxidative stress and cardiovascular injury: part II: animal and human studies. Circulation 108:2034–2040

He J, Kryger M, Zorick F, Conway W, Roth T (1988) Mortality and apnea index in obstructive sleep apnea. Experience in 385 male patients. Chest 94:9–14

Hopps E, Canino B, Calandrino V, Montana M, Lo Presti R, Caimi G (2014) Lipid peroxidation and protein oxidation are related to the severity of OSAS. Eur Rev Med Pharmacol Sci 18:3373–3778

Kapur V, Auckley D, Chowdhuri S, Kuhlmann D, Mehra R, Ramar K, Harrod C (2017) Clinical practice guideline for diagnostic testing for adult obstructive sleep apnea: an American Academy of Sleep Medicine Clinical Practice Guideline. J Clin Sleep Med 13 (3):479–504

Lavie L (2009) Oxidative stress – a unifying paradigm in obstructive sleep apnea and comorbidities. Prog Cardiovasc Dis 51:303–312

Lavie L, Lavie P (2009) Molecular mechanisms of cardio-vascular disease in OSAHS: the oxidative stress link. Eur Respir J 33:1467–1484

Lavie P, Herer P, Peled R, Berger L, Yoffe N, Zomer J, Ami–Hai E (1995) Sleep apnea research mortality in sleep apnea patients: a multivariate analysis of risk factors. Sleep 18:149–157

Marin JM, Carrizo SJ, Vicente E, Agusti AG (2005) Long-term cardiovascular outcomes in men with obstructive sleep apnoea-hypopnoea with or without treatment with continuous positive airway pressure: an observational study. Lancet 365:1046–1053

McNicholas WT, Bonsignore MR; Management Committee of EU Cost Action B26 (2007) Sleep apnoea as an independent risk factor for cardiovascular disease: current evidence, basic mechanisms and research priorities. Eur Respir J 29:156–178

Ntalapascha M, Makris D, Kyparos A, Tsilioni I, Kostikas K, Gourgoulianis K, Kouretas D, Zakynthinos E (2012) Oxidative stress in patients with obstructive sleep apnea syndrome. Sleep Breath 19:549–555

Passali D, Corallo G, Yaremchuk S, Longini M, Proietti F, Passali GC, Bellussi L (2015) Oxidative stress in patients with obstructive sleep apnoea syndrome, vol 35, pp 420–425

Robinson GV, Peprell JC, Segal HC, Davies RJ, Strandling JR (2004) Circulating cardiovascular risk factors in obstructive sleep apnoea: data from randomized controlled trials. Thorax 59:777–782

Sales LV, Bruin VM, D'Almeida V, Pompéia S, Bueno OF, Tufik S, Bittencourt L (2013) Cognition and biomarkers of oxidative stress in obstructive sleep apnea. Clinics (Sao Paulo) 68:449–455

Somers V, White D, Amin R, Abraham W, Costa F, Culebras A, Daniels S, Floras J, Hunt C, Olson L, Pickering T, Russel R, Woo M, Young T (2008) Sleep apnea and cardiovascular disease: an American Heart Association/American College of Cardiology Foundation Scientific Statement from the American Heart Association Council for High Blood Pressure Research Professional Education Committee, Council on Clinical Cardiology, Stroke Council, and Council on Cardiovascular Nursing. J Am Coll Cardiol 52:686–717

Svatikova A, Wolk R, Lerman LO, Juncos LA, Greene EL, McConnell JP, Somers VK (2005) Oxidative stress in obstructive sleep apnoea. Eur Heart J 26:2435–2439

Youseff HA, Elsahzly MI, Rashed LA, Sabry IM, Ibrahim EK (2014) Thiobarbituric acid reactive substances (TBARS) a marker of oxidass in obstructive sleep apnea. Egypt J Chest Dis Tuberc 63:119–124

Advs Exp. Medicine, Biology - Neuroscience and Respiration (2019) 44: 37–42
https://doi.org/10.1007/5584_2019_419
© Springer Nature Switzerland AG 2019
Published online: 1 August 2019

Prevalence and Risk of Obstructive Sleep Apnea and Arterial Hypertension in the Adult Population in Poland: An Observational Subset of the International Prospective Urban Rural Epidemiology (PURE) Study

Katarzyna Postrzech-Adamczyk, Artur Nahorecki, Katarzyna Zatońska, Joshua Lawson, Maria Wołyniec, Robert Skomro, and Andrzej Szuba

Abstract

Obstructive sleep apnea (OSA) is a common breathing disorder affecting millions of people worldwide. The disorder is connected with serious consequences including hypertension, myocardial infarction, arrhythmias, coronary disease, cardiac insufficiency, stroke, transient ischemic attack, and cognitive decline. Epidemiological data assessing the prevalence of OSA in different countries vary in methodology, size, and characteristics of population chosen and thus are hardly comparable. There are very few reports on the prevalence of OSA and on the diagnostic accuracy of sleep questionnaires available in Poland. In this report we present the analysis of the prevalence of, and risk factors for OSA in the Polish adult population consisting of 613 community-based subjects (227 men and 386 women). The study was based on the STOP-BANG questionnaire, a validated Screening Tool for OSA in primary care. It is a part of Polish subset of the Prospective Urban Rural Epidemiology (PURE) study, an ongoing population cohort study of individuals from urban and rural communities from 21 countries. We took into account age, gender, body mass index (BMI), and antihypertensive treatment. The findings are that over one half of the Polish population investigated had a moderate to high risk of OSA (66.5% of men and 60.1% of women). After the adjustments for age, gender, and

K. Postrzech-Adamczyk (✉) and A. Szuba
Angiology Division, Wroclaw Medical University, Wroclaw, Poland

Department of Internal Medicine, Fourth Military Hospital, Wroclaw, Poland
e-mail: kpostrzech.adamczyk@gmail.com

A. Nahorecki
Angiology Division, Wroclaw Medical University, Wroclaw, Poland

K. Zatońska and M. Wołyniec
Department of Social Medicine, Wroclaw Medical University, Wroclaw, Poland

J. Lawson
Center for Health and Safety in Agriculture, University of Saskatchewan, Saskatoon, Canada

R. Skomro
Angiology Division, Wroclaw Medical University, Wroclaw, Poland

University of Saskatchewan, Saskatoon, Canada

BMI we noticed a dose-response relationship between arterial blood pressure behavior and OSA. The association was significant among women, but not men. Based on previous studies we can assume that one half of this high risk group would be further diagnosed for OSA. This study, the first large scale screening for OSA in Poland, shows a substantial, much higher than previously appreciated, prevalence of risk for OSA in the population at large.

Keywords

Adult population · Hypertension · Obstructive sleep apnea · PURE study · Screening questionnaire · Sleep-disordered breathing · STOP-BANG questionnaire

1 Introduction

Obstructive sleep apnea (OSA) syndrome is a common breathing disorder affecting millions of people around the world. It is associated with serious consequences including arterial hypertension (HA), myocardial infarction, arrhythmias, coronary disease, cardiac insufficiency, stroke, transient ischemic attack, and cognitive disorders. A link between OSA and obesity, which is sharply on the rise in the developed countries, makes the increasing prevalence of OSA more probable, along with its health consequences at both individual and societal levels.

The gold standard diagnostic test for OSA is overnight polysomnography which assesses the sleep stages, breathing pattern, blood oxygen saturation, heart rate, and the body responses to episodes of apnea or hypopnea. Although OSA can be present without any clinical symptoms, in most cases it is connected with clinical manifestations, such as excessive daytime sleepiness, decreased concentration, and unrefreshing sleep and fatigue. These symptoms among other factors are the basis for screening questionnaires. A perfect screening test should be characterized by high sensitivity (few false negative results) and high specificity (few false positive results). Although high sensitivity can be achieved, all of the available questionnaires have limited specificity (Kapur et al. 2017). The most commonly used screening tools for OSA in clinical practice are: Epworth Sleepiness Scale (ESS), Berlin Questionnaire (BQ), and STOP-BANG questionnaire. The ESS is the oldest scale among those mentioned above, designed originally to measure the general level of daytime sleepiness. It requires the patient to estimate a likelihood of dozing in 8 different situations. A score greater than 10 points to the probability of OSA. Sensitivity of ESS is estimated at 54%, 47%, and 58% for mild, moderate, and severe OSA and specificity at 65%, 62%, and 60%, respectively. The BQ consists of 11 questions grouped into three categories. The first category includes five questions on snoring, the second category includes three questions on daytime sleepiness, and the third one includes one question on the history of hypertension. In a large meta-analysis, sensitivity and specificity of BQ, depending on OSA severity, assessed from the apnea-hypopnea index (AHI), are as follows: AHI ≥ 5 episodes/h – 76% and 59%, AHI ≥ 15 episodes/h – 77% and 44%, and AHI $\geq$ episodes/h 54% and 38%, respectively (Chiu et al. 2017). The STOP-BANG questionnaire includes eight dichotomous (yes/no) questions related to the clinical features of sleep apnea (snoring, tiredness, noticed apnea, high blood pressure, body mass index (BMI), age, neck circumference, and male gender). The total score ranges from 0 to 8. The available data demonstrate a high sensitivity of 88% in detecting a low-grade OSA (AHI 5–15 episodes/h), 90% in moderate OSA (AHI 16–30 episodes/h), and 93% in severe OSA (AHI > 30 episodes/h). As compared with the other questionnaires, the STOP-BANG has the highest sensitivity. The corresponding specificities are 42%, 36%, and 35% (Boynton et al. 2013; Chung et al. 2008).

The STOP-BANG questionnaire has been chosen for our study whose main objective is to screen for the symptoms of OSA, and therefore to assess the potential prevalence of the disorder in Polish subjects taking part in the Prospective Urban Rural Epidemiology (PURE) study. The PURE study, which has by now been conducted in 21 low-, middle-, and high-income countries examines the influence of societal factors on human lifestyle, cardiovascular disorder risk, and the rate of non-communicable diseases in both urban and rural community inhabitants (Teo et al. 2009).

2   Methods

There were 613 subjects (227 men and 386 women) of the mean age of 59.8 ± 8.8 years (range 29–81 years) who participated in the PURE study in Poland. The subjects were interviewed in the home setting. Each of them underwent a medical examination followed by completion of STOP-BANG questionnaire. The STOP-BANG score was evaluated as a three-category variable: low risk (0–2), medium risk (3–5), and high risk (6–8), and additionally as high vs. low and medium risks combined. All subjects had their height and weight, and blood pressure readings taken at the time of the interview. Blood pressure of more than 130/90 and the use of antihypertensive drugs defined hypertension.

Data were expressed as means ±SD. Following a descriptive analysis, the adjustments for age, gender, and BMI were completed through multiple logistic regression for binary outcomes (e.g., any hypertension) and multiple linear regression for continuous outcomes (e.g., blood pressure). A p-values <0.05 defined statistically significant changes. The analysis was performed using a commercial statistical package of IBM SPSS Statistics (Armonk, NY).

3   Results

Moderate risk of OSA was noticed in 92 (40.5%) men and 179 (46.4%) women and high risk was in 59 men (26.0%) and 53 women (13.7%). The mean score of STOP-BANG questionnaire for the whole cohort of subjects was 4.0 ± 0.9 and 6.5 ± 0.6 in the moderate and high risk OSA groups, respectively. The mean age of men was akin to that of women in both moderate risk (60.7 vs. 60.4, respectively) and high risk OSA (61.6 vs. 61.2, respectively) groups. Likewise, there were no significance inter-gender differences in BMI in both moderate risk (26.9 ± 3.4 kg/m^2 in men vs. and 27.8 ± 5.9 kg/m^2 in in women) and high risk OSA (30.0 ± 4.7 kg/m^2 in men and 31.1 ± 6.8 in women kg/m^2) groups (Table 1). While the overall prevalence of hypertension was 47.6% in the whole cohort of subjects, as the category risk of OSA increased in the STOP-BANG scoring, so did the prevalence of hypertension; it was 25.3% for low, 48.0% for medium, and 68.5% for high risk (p < 0.001) (Fig. 1). The association between the category risk of OSA and the prevalence of hypertension remained statistically significant at p < 0.001 in the whole cohort after adjustments for age, gender, and BMI. In the analysis stratified by gender, the significance of this association was sustained for women (p < 0.001), but not for men (p = 0.92). To complement these results, similar patterns were observed when considering the outcome of systolic blood pressure alone both in the whole cohort

Table 1 Basic demographics and arterial blood pressure in the population cohort screened for obstructive sleep apnea (OSA), stratified by the category risk in the STOP-BANG questionnaire

Category of OSA risk	0–2 (low risk)		3–5 (moderate risk)		6–8 (high risk)	
	Male	Female	Male	Female	Male	Female
Gender	(n = 76)	(n = 154)	(n = 92)	(n = 179)	(n = 59)	(n = 53)
Age (years) SD	56.8 ± 9.0	58.1 ± 8.9	60.7 ± 9.4	60.4 ± 7.2	62.6 ± 7.6	61.2 ± 6.4
BMI (kg/m^2)	26.5 ± 4.2	26.4 ± 4.5	26.9 ± 3.4	27.8 ± 5.9	30.0 ± 4.7	31.1 ± 6.8
STOP-BANG score	1.3 ± 0.8	1.4 ± 0.7	4.1 ± 0.5	3.9 ± 0.5	6.5 ± 0.7	6.5 ± 0.6
Systolic pressure (mmHg)	136.1 ± 12.6	134.8 ± 45.2	141.5 ± 16.4	138.7 ± 19.7	149.0 ± 15.4	146.9 ± 18.5
Diastolic pressure (mmHg)	84.9 ± 9.5	82.7 ± 9.0	85.6 ± 8.9	83.8 ± 10.5	91.0 ± 10.4	87.2 ± 9.6

Data are means ±SD

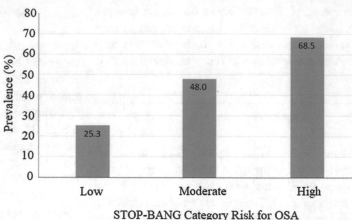

Fig. 1 Prevalence of arterial blood hypertension by the STOP-BANG category risk for obstructive sleep apnea in the population cohort investigated; p < 0.001 for inter-group differnces

and after stratification by gender through crude and adjusted analyses. Although the prevalence of hypertension significantly increased with increasing categories of OSA risk, according to STOP-BANG score, there were no significant inter-categorical differences in the elevated arterial blood pressure in the absolute terms (Table 1).

4   Discussion

This is the first large scale study using STOP-BANG as a validated OSA screening tool in the community-based population cohort in Poland. The available epidemiological data assessing the prevalence of OSA in different countries vary in the methodology, size, and characteristics of group chosen and thus are hardly comparable. Sensitivity of STOP-BANG for scores 3 and above is estimated at about 88% for detecting mild OSA, 90% for moderate OSA, and over 93% for severe OSA (Chiu et al. 2017). In the American population, Peppard et al. (2013) have estimated the incidence of moderate-to-severe OSA (AHI $\geq$15) at 17% among 50–70-year-old men and 9% among 50–70-year-old women. In Poland there are very few reports available on the prevalence of OSA or on the diagnostic accuracy of sleep questionnaires. In a population study on the incidence of OSA among the inhabitants of Warsaw, 676 subjects underwent polysomnography. The mean age of the subjects was 56.6 $\pm$ 8.2 years. OSA (AHI

> 5) was diagnosed in 36.5% of men and 18.5% of women. However, when for the cut-off AHI was taken as >10, the incidence was 19.8% of men and 8.5 of women (Pływaczewski et al. 2008). Kiełbasa et al. (2016) analyzed 261 hypertensive patients (130 women and 131 men) from the outpatient Hypertension Clinic in Cracow. The risk of OSA was evaluated on the basis of STOP-BANG questionnaire. The score above 4 was considered a suspicion of the OSA presence. Seventy three percent of the patients were overweight, with BMI above 25 kg/m^2, and 61% also were diagnosed with metabolic syndrome. About 50% of all those patients scored more than 4 in the STOP-BANG, but in a subgroup with metabolic syndrome the percentage was higher, reaching 55% of patients.

The association between hypertension and OSA has been repeatedly documented. The incidence of OSA in hypertensive patients is estimated at up to 30–50%, but it can be even greater as OSA is often underdiagnosed (Ahmad et al. 2017). The prevalence of hypertension is growing along with the severity of OSA. In a study of Lavie et al. (2000), 53% of patients with severe OSA had hypertension as compared to 46% of those with moderate OSA. In patients suffering from resistant hypertension, the prevalence is even greater, reaching 83% in a study of Logan et al. (2001) performed in 41 patients. In the RESIST-POL study in Poland, 204 patients with resistant hypertension (daytime mean blood pressure > 135/85 mmHg) have been evaluated for OSA, renal artery stenosis, primary aldosteronism, and other secondary forms of hypertension. Mild OSA was present in 27.0%, moderate in 18.6%, and severe in 26.5% patients (Florczak et al. 2013). Risk factors for OSA and hypertension often are mutually overlapping, e.g., male gender, older age, or increased BMI. Nonetheless, epidemiological studies show divergent results regarding the association between OSA and hypertension. In the Wisconsin Sleep Cohort Study patients with AHI 5–15/h and those with AHI >15/h had the odds ratio of having hypertension twofold and threefold greater, respectively, at 4-year-follow-up, compared to those with no episodes of apnea or hypopnea (Peppard et al. 2000). In contradistinction, the Vitoria Sleep Cohort study has shown no significant association between OSA and the incidence of hypertension (Cano-Pumarega et al. 2011). The divergent results can be partially explained by some specific population characteristics, such as age differences, which also has been underlined in other studies. In younger patients (<60 years old), the link OSA-hypertension is more pronounced, while in patients older than 60 years of age, where isolated systolic hypertension is more frequent, this link is no longer that much significant (Haas et al. 2005; Bixler et al. 2000).

5   Conclusions

Based on the present results, over one half of the Polish adult population may be at moderate-to-high risk of OSA (66.5% of men and 60.1% of women). Low specificity of STOP-BANG questionnaire limits its predictive value as a screening tool for OSA. All patients from moderate and high risk groups should be referred to a sleep laboratory, as polysomnography is routinely indicated for the diagnosis of sleep related breathing disorders. Based on the previous studies, we can assume that one half of our high risk group could be diagnosed with OSA. The results also show that after the adjustments for age, gender, and BMI there is a high pretest probability of OSA coexisting with arterial blood hypertension

in a dose-response fashion. This association is significant in women, but not men, which, however, may have to do with a large disproportion between the number of women and men participating in the study.

Acknowledgements Funded, in part, by grants 290/W-PURE/2008/0 from the Polish Ministry of Science and Higher Education and 2015/17/B/NZ7/02963 from the Polish National Science Center. The PURE-Poland study also was partially funded by the Population Health Research Institute in Hamilton, Canada.

Conflicts of Interest The authors declare no conflicts of interest in relation to this article.

Ethical Approval All procedures performed in studies involving human participants were in accordance with the ethical standards of the institutional and/or national research committee and with the 1964 Helsinki declaration and its later amendments or comparable ethical standards. The study was approved by an institutional Ethics Committee.

Informed Consent Written informed consent was obtained from all individual participants included in the study.

References

Ahmad M, Makati D, Akbar S (2017) Review of and updates on hypertension in obstructive sleep apnea. Int J Hypertens 2017(8):1–13

Bixler EO, Vgontzas AN, Lin HM, Ten Have T, Leiby BE, Vela-Bueno A, Kales A (2000) Association of hypertension and sleep-disordered breathing. Arch Intern Med 160(15):2289–2295

Boynton G, Vahabzadeh A, Hammoud S, Ruzicka DL, Chervin RD (2013) Validation of the STOP-BANG questionnaire among patients referred for suspected obstructive sleep apnea. J Sleep Disord Treat Care 2 (4). https://doi.org/10.4172/2325-9639.1000121

Cano-Pumarega I, Durán-Cantolla J, Aizpuru F, Miranda-Serrano E, Rubio R, Martínez-Null C, de Miguel J, Egea C, Cancelo L, Alvarez A, Fernández-Bolaños M, Barbé F (2011) Obstructive sleep apnea and systemic hypertension. Am J Respir Crit Care Med 184 (11):1299–1304

Chiu HY, Chen PY, Chuang LP, Chen NH, Tu YK, Hsieh YJ, Wang YC, Guilleminault CG (2017) Diagnostic accuracy of the Berlin questionnaire, STOP-BANG, STOP, and Epworth sleepiness scale in detecting obstructive sleep apnea: a bivariate meta-analysis. Sleep Med Rev 36(2017):57–70

Chung F, Yegneswaran B, Liao P, Chung SA, Vairavanathan S, Islam S, Khajehdehi A, Shapiro CM (2008) STOP questionnaire: a tool to screen patients for obstructive sleep apnea. Anesthesiology 108(5):812–821

Florczak E, Prejbisz A, Szwench-Pietrasz E, Śliwinski P, Bieleń P, Klisiewicz A, Michałowska I, Warchoł E, Januszewicz M, Kała M, Witkowski A, Więcek A, Narkiewicz K, Somers VK, Januszewicz A (2013) Clinical characteristics of patients with resistant hypertension: the RESIST-POL study. J Hum Hypertens 27 (11):678–685

Haas DC, Foster GL, Nieto FJ, Redline S, Resnick HE, Robbins JA, Young T, Pickering TG (2005) Age-dependent associations between sleep-disordered breathing and hypertension: importance of discriminating between systolic/diastolic hypertension and isolated systolic hypertension in the sleep heart health study. Circulation 111(5):614–621

Kapur VK, Auckley DH, Chowdhuri S, Kuhlmann DC, Mehra R, Ramar K, Harrod CG (2017) Clinical practice guideline for diagnostic testing for adult, obstructive sleep apnea: an American Academy of Sleep Medicine clinical practice guideline. J Clin Sleep Med 13(3):479–504

Kiełbasa G, Stolarz-Skrzypek K, Pawlik A, Łątka M, Drożdż T, Olszewska M, Franczyk A, Czarnecka D (2016) Assessment of sleep disorders among patients with hypertension and coexisting metabolic syndrome. Adv Med Sci 61(2):261–268

Lavie P, Herer P, Hoffstein V (2000) Obstructive sleep apnoea syndrome as a risk factor for hypertension: population study. BMJ 320(7233):479–482

Logan AG, Perlikowski SM, Mente A, Tisler A, Tkacova R, Niroumand M, Leung RS, Bradley TD (2001) High prevalence of unrecognized sleep apnoea in drug-resistant hypertension. J Hypertens 9 (12):2271–2277

Peppard PE, Young T, Palta M, Skatrud J (2000) Prospective study of the association between sleep-disordered breathing and hypertension. N Engl J Med 342 (19):1378–1384

Peppard PE, Young T, Barnet JH, Palta M, Hagen EW, Hla KM (2013) Increased prevalence of sleep-disordered breathing in adults. Am J Epidemiol 177 (9):1006–1014

Pływaczewski R, Bednarek M, Jonczak L, Zieliński J (2008) Sleep-disordered breathing in a middle-aged and older Polish urban population. J Sleep Res 17 (1):73–81

Teo K, Chow CK, Vaz M, Rangarajan S, Yusuf S, PURE Investigators-Writing Group (2009) The Prospective Urban Rural Epidemiology (PURE) study: examining the impact of societal influences on chronic noncommunicable diseases in low-, middle-, and high-income countries. Am Heart J 158(1):1–7

Advs Exp. Medicine, Biology - Neuroscience and Respiration (2019) 44: 43–53
https://doi.org/10.1007/5584_2019_429
© Springer Nature Switzerland AG 2019
Published online: 17 September 2019

Strategies of Coping with Pain: Differences Associated with the Histological Type of Lung Cancer

Jacek Polański, Beata Jankowska-Polańska, Grzegorz Mazur, and Mariusz Chabowski

Abstract

Behavioral cognitive therapy is recommended for lung cancer-related pain. The aim of the study was to analyze the strategies of coping with pain in relation to the histological type of lung cancer. The study included 257 lung cancer patients, divided into small cell lung carcinoma (SCLC) group (n = 72) and non-small cell lung carcinoma (NSCLC) group (n = 185). Pain was evaluated on a visual analog scale (VAS), while pain-coping strategies with the Coping Strategies Questionnaire. The two groups differed concerning the perception of pain – VAS score of 4.8 ± 2.0 in SCLC vs. 4.2 ± 2.0 in NSCLC group (p = 0.003). SCLC patients were less likely to use the active coping strategies, such as increasing behavioral activity ($13.6 + 7.0$ vs. 16.9 ± 6.9; p = 0.001), and pain control (2.5 ± 1.2 vs. 3.4 ± 1.2; p < 0.001), and were lees able to decrease pain (2.4 ± 1.3 vs. 3.5 ± 1.3; p < 0.001). The most common pain-coping strategy in SCLC was praying or hoping, while it consisted of increased behavioral activity and active coping in NSCLC. Correlation analysis for coping strategies and pain intensity showed a negative influence (increased pain) for the following domains: diverting attention (r = 0.264, $\beta = 0.93$); reinterpreting of pain sensations (r = 0.327, $\beta = 0.97$); catastrophizing (r = 0.383, $\beta = 1.11$); ignoring pain sensations (r = 0.306, $\beta = 0.93$), praying or hoping (r = 0.220, $\beta = 0.76$), coping self-statements (r = 0.358, $\beta = 1.10$), and increased behavioral activity (r = 0.159, $\beta = 0.57$). For pain control (r = -0.423, $\beta = -0.27$) and the ability to decrease pain (r = -0.359, $\beta = -0.27$), a positive influence (decreased pain) was found. The significant independent determinants of pain perception in the NSCLC group were: pain control strategy ($\beta = -0.39$) and coping self-statement ($\beta = 0.72$). We conclude that in NSCLC patients both pain control and the ability to decrease pain are the strategies which decrease

J. Polański
Lower Silesian Oncology Center, Home Hospice, Wroclaw, Poland

B. Jankowska-Polańska (✉)
Department of Clinical Nursing, Faculty of Health Science, Wroclaw Medical University, Wroclaw, Poland
e-mail: bianko@poczta.onet.pl

G. Mazur
Department of Internal Medicine, Occupational Diseases, Hypertension and Clinical Oncology, Wroclaw Medical University, Wroclaw, Poland

M. Chabowski
Department of Surgery, Fourth Military Teaching Hospital, Wroclaw, Poland

Division of Surgical Procedures, Department of Clinical Nursing, Faculty of Health Science, Wroclaw Medical University, Wroclaw, Poland the intensity of perceived pain. In contrast, SCLC patients have no clear strategy for pain-coping.

Keywords

Cognitive therapy · Coping with pain · Histology · Lung cancer · Pain control

1   Introduction

Lung cancer is classified according to histological type, which plays a crucial role for therapy and for the anticipation of long-term results. There are two main types of lung cancer. Small-cell lung cancer (SCLC), which represents fast growth, high malignancy, early dissemination, and an advanced disease at the diagnosis in the majority of the patients (70%). This type of lung cancer is not suitable for surgery and only 12–25% of cases are treated with chemo-radiotherapy (Travis et al. 1995). The other type is non-small-cell lung cancer (NSCLC) which constitutes about 80% of all cases and which is the predominant cause of death from lung cancer worldwide, approx. 1,370,000 a year (Chouaid et al. 2013). NSCLC is treated mainly with surgery and radiotherapy, but rarely with chemotherapy.

Recently, patients with lung cancer are treated much more often than it was feasible two or more decades ago. Treatment decisions often take into account specific histologic and genetic characteristics of a tumor. Yet the improvement in five-year survival has not been achieved. The quality of life has become the essential endpoint in the evaluation of therapy. Mood and strategies for coping with negative emotions are the factors being evaluated in oncological patients. The life experience of patients and their sensitivity to stimuli affect the perception of disease and its symptoms. The ability of coping with cancer may underlie variability in the patient's perception of information, symptoms, and therapy (Chabowski et al. 2018). This ability of coping has been considered a trait acquired at an early period of the disease. Antonovsky (1979) defines

the coping ability/sense of coherence as "global orientation that expresses the extent to which one has a pervasive, enduring though dynamic feeling of confidence that the stimuli deriving from one's internal and external environments in the course of living are structured, predictable, and explicable".

In 1979, the International Association for the Study of Pain and the Taxonomy Working Group (1986) defined the pain as "an unpleasant sensory and emotional experience associated with actual or potential tissue damage, or described in terms of such damage". Each individual percepts the pain stimulus in a subjective and unique way. The psychological response to pain are anxiety and fear. Pain is the commonest symptom of a neoplastic disease. It occurs in 25–50% of patients with lung cancer as a result of the pleura or thoracic wall invasion, pulmonary embolism, or pneumonia. Pain might be caused by the diagnostic and treatment procedures or might be connected with some symptoms such as fatigue and depression. The intensity of pain depends on the type of cancer, its histological features, its localization, treatment method, and co-existing diseases. There are only a few papers focusing on the association between the intensity of pain and psychological and emotional factors (Simmons et al. 2012). Due to a multifactorial origin of pain, a multidisciplinary approach for its treatment is needed.

A strategy of coping with pain is an important element of its effective treatment. It refers to the way which individuals, who experiences pain, develop to tolerate, minimize, or reduce pain (Jensen et al. 1991). The influence of the coping style with pain and adjustments to it are believed to be crucial. However, some cognitive strategies, such as "catastrophizing" or "praying or hoping" may be maladaptive and strongly associated with poor adjustment and unfavorable results. Taking into account the pain-coping strategy may contribute to the understanding of pain determinants and their influence on the patient's attitude toward both disease and therapy. Therefore, this study was undertaken to define the pain-coping strategies depending on the histological type and treatment method in patients with lung cancer. We attempted to show a possible association between the intensity of pain and the coping strategy. The following two hypotheses were made: (1) negative pain-coping strategies would predominate in patients with lung cancer regardless of its histological type and (2) positive pain-coping would decrease pain intensity.

2 Methods

2.1 Patients

The study group consisted of 257 patients (115 women and 142 men; mean age 63.2 ± 9.4). They were divided according to histologic subtype into small cell lung carcinoma (SCLC) (n = 72; mean age 64.1 ± 8.8, range 31–81 years) and non-small cell lung carcinoma (NSCLC) (n = 185; mean age 62.8 ± 9.6, range 25–87 years) with. All patients were treated in the Lower Silesian Centre for Pulmonary Diseases in Wroclaw, Poland, between January and December 2015. Patients filled out the questionnaire. The sociodemographic data were obtained from medical files.

Inclusion criteria were as follows:

- pathological and firm diagnosis of lung cancer
- age > 18 years
- agreement for participation in the study
- understanding of the questions in the questionnaire

Patients with serious co-morbidities, which could disturb the perception of one's health, such as another neoplasm, cardiac failure, severe COPD or asthma, cardiovascular instability, and cognitive impairment, were excluded from the study.

2.2 Measurement Tools

The survey method was used in the study. All patients received questionnaires, along with written information on the study and anonymity of responses. Sociodemographic data on gender,

age, marital status, education, and livelihoods were collected. Then, patients completed the questionnaires with the assistance of an investigator.

Pain Coping Strategies Questionnaire (CSQ)
This is a 48-item tool developed by Rosenstiel and Keefe (1983), with later modifications (Swartzman et al. 1994), which was used in a Polish adaptation by Juczynski (2001). The questionnaire contains 48 items related to six cognitive types of coping strategies: diverting attention ("I do something I enjoy, such as watching TV or listening to music"); reinterpreting pain sensations ("I imagine that the pain is outside of my body"); catastrophizing ("It is terrible and I feel it is never going to get any better"); ignoring pain sensations ("I do not think about the pain"); praying or hoping ("I pray for the pain to stop"); coping self-statements ("I see it as a challenge and do not let it bother me"), and two behavioral strategies: increasing activity level and increasing pain behaviors ("I do anything to get my mind off the pain"). The questionnaire ends up with two items that concern the perceived effectiveness of coping: control over pain and the ability to decrease pain. The strategies above outlined are grouped into three scales: active and generally adaptive coping (reinterpreting pain sensations, coping self-statements, and ignoring pain sensations); distancing from pain (diverting attention and increasing activity level); and passive and maladaptive coping (catastrophizing and praying or hoping). Patients rate the frequency of their use of a specific strategies on a seven-point Likert-type scale ranging from 0 "never" to 6 "always". Each domain is scored separately, with a higher score indicating a greater use. Standardized Cronbach's α coefficient ranges from 0.63 (diverting attention) to 0.82 (catastrophizing, coping self-statements), indicating an acceptable reliability of the questionnaire.

Zubrod/ECOG (Eastern Cooperative Oncology Group) Performance Status Scale This scale, was published by Oken et al. (1982), marks a functional status of oncological patients, ranging

Table 1 Sociodemographic features of small cell lung cancer (SCLC) and non-small cell lung cancer (NSCLC) patients

Variable	Total	Histological type		SCLC vs. NSCLC
		SCLC	NSCLC	
Patient counts; n (%)	257	72 (28)	185 (72)	p-value
Age (year)				0.357
Mean ± SD	63.2 ± 9.4	64.1 ± 8.8	62.8 ± 9.6	
Min–Max	25–87	31–81	25–87	
Place of living; n (%)				0.160
Urban area	179 (69.6)	45 (62.5)	134 (72.4)	
Rural area	78 (30.4)	27 (37.5)	51 (27.6)	
Source of income n (%)				0.264
Permanent job	68 (26.5)	14 (19.4)	54 (29.2)	
Pension/Retirement pension/Dole	181 (70.4)	56 (77.8)	125 (67.6)	
Relative-dependent	8 (3.1)	2 (2.8)	6 (3.2)	
Education; n (%)				0.054
Primary	29 (11.3)	14 (19.4)	15 (8.1)	
Vocational	118 (45.9)	28 (38.9)	90 (48.6)	
Secondary	86 (33.5)	25 (34.7)	61 (33.0)	
Higher	24 (9.3)	5 (6.9)	19 (10.3)	
Cancer in family; n (%)				0.990
Yes	98 (38.1)	28 (38.9)	70 (37.8)	
No	159 (61.9)	44 (61.1)	115 (62.2)	

from fully functional – 0 points to disability and confinement to bed/chair – 5 points. The intensity of pain was measured on a 10-point visual analog scale (VAS). The score ranges from 0 to 10 depending on the severity of pain, with the extreme points marking "no pain" and "the worst possible pain" (Hawker et al. 2011).

2.3 Statistical Evaluation

Data were expressed as means ±SD and medians in case of non-normal distribution, which was checked with the Kolmogorov-Smirnov test. Differences between the SCLC and NSCLC groups was tested with Student's *t*-test. In case of non-normal distribution of data, the Mann–Whitney U test was used. Pearson's chi-squared test ($\chi 2$) was applied to sets of ordinal or categorical data to evaluate statistical differences between them. A p-value <0.05 defined statistically significant differences.

3 Results

3.1 Sociodemographic Features of Small Cell Lung Cancer (SCLC) and Non-Small Cell Lung Cancer (NSCLC) Patients

There were no significant differences in the sociodemographic features between the SCLC and NSCLC patients. The majority of patients lived in urban areas (69.9%), had vocational (45.9%) or high school education (33.5%), and were pensioners (70.4%). The families of 61.9% patients with lung cancer were free from other cancer diseases (Table 1).

3.2 Clinical Features in Small Cell Lung Cancer (SCLC) and Non-small Cell Lung Cancer (NSCLC) Patients

There were differences between SCLC and NSCLC groups concerning the involvement of regional lymph nodes (N) and distant metastases

(M) (Travis et al. 2015). In the NSCLC group, mediastinal lymph nodes (N0) were involved significantly less often (36.8% vs. 26.4% in SCLC; p = 0.015) whereas in the SCLC group, ipsilateral mediastinal (N2) (33.3% vs. 28.1% in NSCLC; p = 0.015), or contralateral mediastinal, or supraclavicular lymph nodes (N3) (6.9% vs 5.9% in NSCLC; p = 0.015) were involved significantly more often. Likewise, in the NSCLC group, distant metastases was revealed less often (M0) (69.2% vs. 44.4% in SCLC; p = 0.004), whereas in the SCLC group, metastases in the chest (M1) (27.8% vs. 18.4% in NSCLC; p = 0.004) or outside of the chest (M2) (4.2% vs. 2.2% in NSCLC; p = 0.004) or of unknown localization (Mx) (22.2% vs. 9.2% in NSCLC; p = 0.004) were more often confirmed. Patients with SCLC had more often metastases to liver (29.2% vs. 9.2% in NSCLC; p < 0.001), were more often admitted to the hospital (2.7 ± 2.7 vs. 1.0 ± 1.7 in NSCLC; p < 0.001), and more often smoked cigarettes (55.6% vs. 37.3% in NSCLC; p = 0.005). As regards the symptoms, there were no significant differences. Chronic cough (90.3% vs. 80.5%; p = 0.091), dyspnea (73.6% vs. 61.6%, p = 0.096), and chest pain (54.2% vs. 40.0%; p = 0.056) were most often revealed in the SCLC and NSCLC groups, respectively (Table 2).

As regards the methods of treatment, SCLC patients less often underwent surgery, but more often had chemotherapy and supportive care (26.4% vs. 8.6%; p < 0.001). They also had significantly worse forced expiratory volume in 1 s (FEV1) and forced vital capacity (FVC), and had more intense pain measured by on VAS, compared with NSCLC patients. There were no appreciable differences in performance status between the two groups of patients. Detailed values are displayed in Table 3.

3.3 Strategies for Coping with Pain in Small Cell Lung Cancer (SCLC) and Non-small Cell Lung Cancer (NSCLC) Patients

The SCLC patients scored lower on CSQ. They used less often the positively connoted strategies such as increasing behavioral activity, pain control, and the ability to decrease pain perception. On the other side, they more often than the NSCLC patients used the praying and hoping strategy. Details of pain-coping strategies displayed by both SCLC and NSCLC patients are shown in Table 4.

3.3.1 Association Between Strategies for Coping with Pain and Intensity of Pain

The Entire Study Group – SCLC and NSCLC Patients Combined (n = 257) There were significant associations between the strategies for coping with pain and the intensity of pain in the univariate and multivariate regression analyses. The following domains: diverting attention ($r = 0.264$, $\beta = 0.93$), reinterpreting of pain sensations ($r = 0.327$, $\beta = 0.97$), catastrophizing ($r = 0.383$, $\beta = 1.11$), ignoring pain sensations ($r = 0.306$, $\beta = 0.93$), praying or hoping ($r = 0.220$, $\beta = 0.76$), coping self-statements ($r = 0.358$, $\beta = 1.10$), and increasing behavioral activity ($r = 0.159$, $\beta = 0.57$) positively associated with pain, meaning they increased pain intensity. In contrast, the domains pointing to the subject's independent attitude, i.e., pain control ($r = -0.423$, $\beta = -0.27$) and the ability to decrease pain ($r = -0.359$, $\beta = -0.27$), inversely associated with pain, meaning they decreased pain intensity (Table 5).

Small Cell Lung Cancer (SCLC) Patients (n = 72) Significant associations between the strategies for pain-coping and the intensity of pain in the SCLC patients, which increase the perception of pain, were revealed only in the univariate regression analysis and concerned the domains of reinterpreting of pain sensations ($r = 0.389$), catastrophizing ($r = 0.470$), ignoring pain sensations ($r = 0.314$), and coping self-statements ($r = 0.287$). There was no inverse association revealed, which could help decrease pain intensity (Table 6).

Non-small Cell Lung Cancer (NSCLC) Patients (n = 185) In the univariate regression analysis, all strategies coping with pain, except pain control and the ability to decrease pain, associated

Table 2 Clinical features of small cell lung cancer (SCLC) and non-small cell lung cancer (NSCLC) patients

Variable	Total	Histological type		SCLC vs. NSCLC
		SCLC	NSCLC	
Patient counts; n (%)	257	72 (28)	185 (72)	p-value
TNM:				0.357
T1	48 (18.7)	7 (9.7)	41 (22.2)	
T2	86 (33.5)	10 (13.9)	76 (41.1)	
T3	29 (11.3)	8 (11.1)	21 (11.4)	
T4	90 (35.0)	44 (61.1)	46 (24.9)	
Tx	2 (0.8)	0 (0.0)	2 (1.1)	
N0	87 (33.9)	19 (26.4)	68 (36.8)	**0.015**
N1	52 (20.2)	10 (13.9)	42 (22.7)	
N2	76 (29.6)	24 (33.3)	52 (28.1)	
N3	16 (6.2)	5 (6.9)	11 (5.9)	
Nx	24 (9.3)	13 (18.1)	11 (5.9)	
M0	160 (62.3)	32 (44.4)	128 (69.2)	**0.004**
M1	54 (21.0)	20 (27.8)	34 (18.4)	
M2	7 (2.7)	3 (4.2)	4 (2.2)	
M3	1 (0.4)	0 (0.0)	1 (0.5)	
Mx	33 (12.8)	16 (22.2)	17 (9.2)	
Number of hospitalizations				**<0.001**
Mean ± SD	1.5 ± 2.1	2.7 ± 2.7	1.0 ± 1.7	
Range	0–11	0–11	0–11	
Smoking habit				**0.005**
Yes	109 (42.4)	40 (55.6)	69 (37.3)	
No, never	41 (16.0)	5 (6.9)	36 (19.5)	
Quit smoking	91 (35.4)	20 (27.8)	71 (38.4)	
Passive smoker	16 (6.2)	7 (9.7)	9 (4.9)	
Chronic diseases				
Diabetes	75 (29.2)	21 (29.2)	54 (29.2)	0.881
Ischemic heart disease	47 (18.3)	19 (26.4)	28 (15.1)	0.055
Renal insufficiency	11 (4.3)	5 (6.9)	6 (3.2)	0.330
Rheumatoid arthritis	14 (5.4)	10 (13.9)	4 (2.2)	**0.001**
Cardiac failure	59 (23.0)	23 (31.9)	36 (19.5)	**0.049**
Asthma/COPD	59 (23.0)	21 (29.2)	38 (20.5)	0.190
Metastases				
No	156 (60.7)	35 (48.6)	121 (65.4)	**0.020**
Bone	18 (7.0)	8 (11.1)	10 (5.4)	0.181
Brain	16 (6.2)	8 (11.1)	8 (4.3)	0.083
Liver	38 (14.8)	21 (29.2)	17 (9.2)	**<0.001**
Suprarenal	30 (11.7)	10 (13.9)	20 (10.8)	0.636
Multiple	20 (7.8)	4 (5.6)	16 (8.6)	0.604
Symptoms				
Chronic cough	214 (83.3)	65 (90.3)	149 (80.5)	0.091
Dyspnea	167 (65.0)	53 (73.6)	114 (61.6)	0.096
Chest pain	113 (44.0)	39 (54.2)	74 (40.0)	0.056
Hemoptoe	76 (29.6)	19 (26.4)	57 (30.8)	0.585
Recurrent infections	65 (25.3)	13 (18.1)	52 (28.1)	0.132
Vena cava superior syndrome	7 (2.7)	3 (4.2)	4 (2.2)	0.404
Arrhythmia	14 (5.4)	7 (9.7)	7 (3.8)	0.115
Hoarseness	66 (25.7)	14 (19.4)	52 (28.1)	0.205

Table 3 Treatment, pulmonary function, and performance of small cell lung cancer (SCLC) and non-small cell lung cancer (NSCLC) patients

| Variable | Total | Histological type | | SCLC vs. NSCLC |
		SCLC	NSCLC	
Patient counts; n (%)	257	72 (28)	185 (72)	p-value
Methods of treatment				
Surgery	149 (58.0)	17 (23.6)	132 (71.4)	**<0.001**
Radiotherapy	82 (31.9)	27 (37.5)	55 (29.7)	0.293
Chemotherapy	164 (63.8)	62 (86.1)	102 (55.1)	**<0.001**
Supportive care	35 (13.6)	19 (26.4)	16 (8.6)	**<0.001**
FEV1 (L)				**0.014**
Mean ± SD	2.30 ± 0.76	2.12 ± 0.61	2.37 ± 0.80	
FVC (L)				**0.005**
Mean ± SD	2.97 ± 0.95	2.76 ± 0.90	3.06 ± 0.96	
Pain intensity (VAS; score)				**0.003**
Mean ± SD	4.3 ± 2.0	4.8 ± 2.0	4.2 ± 2.0	
Range	0–10	0–8	0–10	
Performance status				0.282
0 – asymptomatic and fully active	47 (18.3)	13 (18.1)	34 (18.4)	
1 – symptomatic but restricted in activity, able to carry out light work	107 (41.6)	27 (37.5)	80 (43.2)	
2 – all self-care but unable to carry out any work, <50% in bed	87 (33.9)	26 (36.1)	61 (33.0)	
3 – limited self-care, >50% in bed	12 (4.7)	3 (4.2)	9 (4.9)	
4 – completely disabled, cannot carry on any self-care	4 (1.6)	3 (4.2)	1 (0.5)	

FEV1 forced expiratory volume in 1 s, *FVC* forced vital capacity; VAS, visual-analog scale

Table 4 Strategies for coping with pain assessed with the Coping Strategies Questionnaire (CSQ) in small cell lung cancer (SCLC) and non-small cell lung cancer (NSCLC) patients

| Pain-coping strategy | Histological type | | p-value |
| | SCLC | NSCLC | |
	n = 72	n = 185	
Diverting attention			0.095
Mean ± SD	14.9 ± 7.1	16.6 ± 7.0	
Median (Q_1; Q_3)	15 (10; 20)	18 (12; 21)	
Reinterpreting of pain sensations			0.484
Mean ± SD	12.7 ± 5.5	13.1 ± 6.1	
Median (Q_1; Q_3)	13 (9; 17)	15 (9; 17)	
Catastrophizing			0.074
Mean ± SD	14.1 ± 5.8	12.9 ± 5.7	
Median (Q_1; Q_3)	15 (11; 19)	14 (9; 17)	
Ignoring pain sensations			0.439
Mean ± SD	12.4 ± 6.4	13.2 ± 6.0	
Median (Q_1; Q_3)	14 (7; 17)	14 (9; 16)	
Praying or hoping			**0.005**
Mean ± SD	19.4 ± 6.3	16.5 ± 6.9	
Median (Q_1; Q_3)	20 (15; 25)	17 (13; 22)	
Coping self-statements			0.134
Mean ± SD	14.6 ± 5.4	15.6 ± 6.4	
Median (Q_1; Q_3)	15 (11; 19)	17 (12; 20)	

(continued)

Table 4 (continued)

Pain-coping strategy	Histological type		p-value
	SCLC n = 72	NSCLC n = 185	
Increasing behavioral activity			**0.001**
M ± SD	13.6 ± 7.0	16.9 ± 6.9	
Median (Q$_1$; Q$_3$)	16 (7; 19)	18 (14; 21)	
Pain control			**<0.001**
M ± SD	2.5 ± 1.2	3.4 ± 1.2	
Median (Q$_1$; Q$_3$)	3 (2; 3)	3 (3; 4)	
Ability to decrease pain			**<0.001**
Mean ± SD	2.4 ± 1.3	3.5 ± 1.3	
Median (Q$_1$; Q$_3$)	2 (2; 3)	3 (3; 4)	

Table 5 Associations between pain intensity, assessed on visual analog scale (VAS), and strategies for coping with pain, assessed with the Coping Strategies Questionnaire (CSQ), in the entire group of lung cancer patients

Strategies for coping with pain	Intensity of pain (VAS)			
	Univariate analysis		Multivariate analysis	
	r	p-value	β	p-value
Diverting attention	0.264	**<0.001**	0.93	**<0.001**
Reinterpreting of pain sensations	0.327	**<0.001**	0.97	**<0.001**
Catastrophizing	0.383	**<0.001**	1.11	**<0.001**
Ignoring pain sensations	0.306	**<0.001**	0.93	**<0.001**
Praying or hoping	0.220	**<0.001**	0.76	**<0.001**
Coping self-statements	0.358	**<0.001**	1.10	**<0.001**
Increasing behavioral activity	0.159	**0.011**	0.57	**0.011**
Pain control	−0.423	**<0.001**	−0.27	**<0.001**
Ability to decrease pain	−0.395	**<0.001**	−0.27	**<0.001**

r correlation coefficient, β regression coefficient

Table 6 Associations between pain intensity, assessed on visual analog scale (VAS), and strategies for coping with pain, assessed with the Coping Strategies Questionnaire (CSQ), in small cell lung cancer (SCLC) patients

Strategies for coping with pain	Intensity of pain (VAS)			
	Univariate analysis		Multivariate analysis	
	r	p-value	β	p-value
Diverting attention	0.182	0.126	0	NS
Reinterpreting of pain sensations	0.389	**<0.001**	0	NS
Catastrophizing	0.470	**<0.001**	0	NS
Ignoring pain sensations	0.314	**0.007**	0	NS
Praying or hoping	−0.128	0.284	0	NS
Coping self-statements	0.287	**0.015**	0	NS
Increasing behavioral activity	0.090	0.451	0	NS
Pain control	−0.188	0.113	0	NS
Ability to decrease pain	−0.155	0.194	0	NS

r correlation coefficient, β regression coefficient

Table 7 Associations between pain intensity, assessed on visual analog scale (VAS), and strategies for coping with pain, assessed with the Coping Strategies Questionnaire (CSQ), in non-small lung cancer (NSCLC) patients

Strategies for coping with pain	Intensity of pain (VAS)			
	Univariate analysis		Multivariate analysis	
	r	p-value	β	p-value
Diverting attention	0.091	**<0.001**	0	NS
Reinterpreting of pain sensations	0.103	**<0.001**	0	NS
Catastrophizing	0.116	**<0.001**	0	NS
Ignoring pain sensations	0.105	**<0.001**	0	NS
Praying or hoping	0.090	**<0.001**	0	NS
Coping self-statements	0.124	**<0.001**	0.07	**<0.001**
Increasing behavioral activity	0.067	**0.001**	0	NS
Pain control	−0.769	**<0.001**	−0.39	**<0.001**
Ability to decrease pain	−0.675	**<0.001**	0	NS

r correlation coefficient, β regression coefficient

with pain intensity; meaning they increased pain intensity. In the multivariate analysis, such strategies were limited only to coping self-statements ($\beta = 0.07$; $p < 0.001$). The domain pain control inversely associated with pain intensity ($\beta = -0.39$; $p < 0.001$), meaning it decreased the perception of pain (Table 7).

4　Discussion

This study revealed that SCLC patients, who admitted in more advanced stages of cancer and thus often with metastases to regional lymph nodes and distant organs, more often experienced pain compared with NSCLC patients. In fact, SCLC patients were more often treated with combined therapy, which is in line with current recommendations. Such treatment is more aggressive and associated with more toxicity. These findings are in line with a study by Herndon et al. (1999) who have examined 206 patients with advanced NSCLC and also conclude that the patient-provided pain report has the greatest prognostic importance for survival. Likewise, Iyer et al. (2013) have reported in a multicenter study that over 90% of patients with advanced lung cancer experience pain, which has a negative influence on quality of life. This observation should encourage health care professionals for proper pain management.

The ability of coping with pain is an attempt to change the style of perceiving and thinking about pain (Esteve et al. 2007; Felton et al. 1984). According to Lazarus and Folkman (1984), coping with pain is understood as an effort to adapt to pain or manage the anxiety arising in response to pain. There are two types of pain-coping strategies, i.e., a cognitive response that involves the use of imagination to ignore pain or transform it into another sensation, and a behavioral response that seeks social support, rest, or avoiding some activities or situations (Rodríguez Franco et al. 2004). Gil et al. (1989) have found that passive coping (catastrophizing) is maladaptive and associated with an increase in pain intensity and frequency. In contrast, active coping (ignoring pain sensations and self-verbalization) is adaptive in that it leads to pain decrease. The findings of this study provide information on coping with pain by patients with different histological types of lung cancer. The most common strategy for pain-coping in SCLC was "praying or hoping", whereas it consisted of increased behavioral activity and active coping ("pain control" and "ability to decrease pain") in NSCLC. The "praying or hoping" used by SCLC patients might stem from aggressive oncological treatment and numerous hospitalizations, during which health control rests with medical professionals.

There are scarce publications focusing on coping strategy with cancer pain and its influence on quality of life. Porter et al. (2011) have suggested in a study on 233 patients with lung cancer that psychosocial interventions can lead to improvements in pain perception, depression, quality of life, and self-efficacy. Liao et al. (2014) have shown in a study on 101 patients with newly diagnosed advanced lung cancer that self-efficacy for coping with cancer is the most important factor for predicting quality of life. In the present study, we noticed a significant association between strategies for pain-coping and pain intensity. Taking into account all patients, irrespective of the type of lung cancer, pain caused increases in behavioral and passive strategies, but decreased coping self-statements. In NSCLC patients, pain positively influenced coping self-statements but inversely affected pain control. Porter et al. (2008) have examined self-efficacy for managing pain in 152 patients with early stage lung cancer and concluded that interventions targeted at increasing self-efficacy may be useful in this population. Thus, therapeutic interventions, such as training of cognitive pain-coping skills, may be beneficial in pain management and in reduction of psychological distress in lung cancer patients.

This study has some limitations such as a heterogenous group of patients, having different co-morbidities and being at different stages of lung cancer; a potential bias arising from the use of self-reported measures; and a lack of items referring to direct coping with pain, i.e., taking medications or consultation with a doctor, in the CSQ questionnaire. There was no control group either, so that associations were identified only among lung cancer patients. Despite these constraints, we believe we have shown the presence of associations between distinct strategies for pain-coping and the histological subtype of lung cancer. A better understanding of pain-coping strategies among lung cancer patients may help physicians to introduce new interventions to improve therapy outcome.

In conclusion, the most common strategy for coping with pain in SCLC patients was "praying or hoping", a strategy that fails to positively influence pain perception. In contrast, in NSCLC patients coping with pain consisted of increased behavioral activity and active coping, the strategies that are advantageous concerning the ability to control the perception of pain intensity in a positive manner. These strategies appeared as independent factors associated with decreased pain intensity.

Conflict of Interests The authors declare no conflicts of interest in relation to this article.

Ethical Approval All procedures performed in studies involving human participants were in accordance with the ethical standards of the institutional and/or national research committee and with the 1964 Helsinki declaration and its later amendments or comparable ethical standards. The study was approved by the Bioethics Committee of Wroclaw Medical University in Wroclaw (permit 507/2015).

Informed Consent Written informed consent was obtained from all individual participants included in the study. In addition, all participants agreed in writing to fill out the questionnaires in the presence of one of the authors of this article.

References

Antonovsky A (1979) Health, stress and coping: new perspectives on mental and physical well-being. Jossey-Bass, San Francisco, p 10

Chabowski M, Jankowska-Polańska B, Lomper K, Janczak D (2018) The effect of coping strategy on quality of life in patients with NSCLC. Cancer Manag Res 10:4085–4093

Chouaid C, Agulnik J, Goker E, Herder GJ, Lester JF, Vansteenkiste J, Finnern HW, Lungershausen J, Eriksson J, Kim K, Mitchell PL (2013) Health-related quality of life and utility in patients with advanced non-small-cell lung cancer: a prospective cross-sectional patient survey in a real-world setting. J Thorac Oncol 8(8):997–1003

Esteve R, Ramirez-Maestre C, Lopez-Marinez AE (2007) Adjustment to chronic pain: the role of pain acceptance, coping strategies, and pain related cognitions. Ann Behav Med 33(2):179–188

Felton BJ, Revenson TA, Hinrichsen GA (1984) Strategies and coping in the explanation of psychological adjustment among chronically ill adults. Soc Sci Med 18 (10):889–898

Gil KM, Abrams MR, Phillips G, Keefe FJ (1989) Sickle cell disease pain: relation of coping strategies to adjustment. J Consult Clin Psychol 57(6):725–731

Hawker GA, Mian S, Kendzerska T, French M (2011) Measures of adult pain: Visual Analog Scale for Pain (VAS Pain), Numeric Rating Scale for Pain (NRS Pain), McGill Pain Questionnaire (MPQ), Short-Form McGill Pain Questionnaire (SF-MPQ), Chronic Pain Grade Scale (CPGS), Short Form-36 Bodily Pain Scale (SF-36 BPS), and Measure of Intermittent and Constant Osteoarthritis Pain (ICOAP). Arthritis Care Res (Hoboken) 63(Suppl 11):S240–S252

Herndon JE 2nd, Fleishman S, Kornblith AB, Kosty M, Green MR, Holland J (1999) Is quality of life predictive of the survival of patients with advanced non-small cell lung carcinoma? Cancer 85(2):333–340

International Association for the Study of Pain, Subcommittee on Taxonomy (1986) Classification of chronic pain. Descriptions of chronic pain syndromes and definitions of pain terms. Pain 3:S1–S226

Iyer S, Taylor-Stokes G, Roughley A (2013) Symptom burden and quality of life in advanced non-small cell lung cancer patients in France and Germany. Lung Cancer 81(2):288–293

Jensen MP, Turner JA, Romano JM, Karoly P (1991) Coping with chronic pain: a critical review of the literature. Pain 47(3):249–283

Juczynski Z (2001) Measurement tools in the promotion and psychology of health. Chapter D: Pain coping strategy questionnaire – CSQ. Laboratory of Psychological Tests of the Polish Psychological Society, Warsaw, pp 162–167 (Article in Polish)

Lazarus RS, Folkman S (1984) Stress, appraisal, and coping. Springer, New York

Liao YC, Shun SC, Liao WY, Yu CJ, Yang PC, Lai YH (2014) Quality of life and related factors in patients with newly diagnosed advanced lung cancer: a longitudinal study. Oncol Nurs Forum 41(2):E44–E55

Oken MM, Creech RH, Tormey DC, Horton J, Davis TE, McFadden ET, Carbone PP (1982) Toxicity and response criteria of the eastern cooperative oncology group. Am J Clin Oncol 5(6):649–655

Porter LS, Keefe FJ, Garst J, McBride CM, Baucom D (2008) Self-efficacy for managing pain, symptoms, and function in patients with lung cancer and their informal caregivers: associations with symptoms and distress. Pain 137(2):306–315

Porter LS, Keefe FJ, Garst J, Baucom DH, McBride CM, McKee DC, Sutton L, Carson K, Knowles V, Rumble M, Scipio C (2011) Caregiver-assisted coping skills training for lung cancer: results of a randomized clinical trial. J Pain Symptom Manag 41(1):1–13

Rodríguez Franco L, Cano García F, Blanco Picabia A (2004) Assessment of chronic pain coping strategies. Actas Esp Psiquiatr 32(2):82–91. (Article in Spanish)

Rosenstiel AK, Keefe FJ (1983) The use of coping strategies in chronic low back pain patients: relationship to patient characteristics and current adjustment. Pain 17(1):33–44

Simmons CP, MacLeod N, Laird BJ (2012) Clinical management of pain in advanced lung cancer. Clin Med Insights Oncol 6:331–346

Swartzman LC, Gwadry FG, Shapiro AP, Teasell RW (1994) The factor structure of the coping strategies questionnaire. Pain 57(3):311–316

Travis WD, Travis LB, Devesa SS (1995) Lung cancer. Cancer 75(1 Suppl):191–202

Travis WD, Brambilla E, Nicholson AG, Yatabe Y, Austin JHM, Beasley MB, Chirieac LR, Dacic S, Duhig E, Flieder DB, Geisinger K, Hirsch FR, Ishikawa Y, Kerr KM, Noguchi M, Pelosi G, Powell CA, Tsao MS, Wistuba I, Panel WHO (2015) The 2015 World Health Organization classification of lung tumors. J Thorac Oncol 10(9):1243–1260

Advs Exp. Medicine, Biology - Neuroscience and Respiration (2019) 44: 55–62
https://doi.org/10.1007/5584_2019_435
© Springer Nature Switzerland AG 2019
Published online: 17 September 2019

Chest Radiography in Children Hospitalized with Bronchiolitis

August Wrotek, Małgorzata Czajkowska,
and Teresa Jackowska

Abstract

In uncomplicated bronchiolitis, chest radiography (CR) is not routinely recommended, yet it is still frequently made. This study seeks to evaluate the use of CR in children with bronchiolitis due to a lower respiratory tract infection (RSV-RTI) with respiratory syncytial virus (RSV) and the influence of CR on patient treatment during the 2010–2017 seasons. There were 581 children included into the study: 459 with bronchiolitis (390 RSV-RTI and 69 non-RSV), 65 with RSV pneumonia and 57 with RSV bronchitis. We found that CR was performed in 28.6% (166/581) patients. CR was much more frequent in patients with RSV than non-RSV infections (61% vs. 31%). CR prognostic sensitivity and specificity in guiding antibiotic treatment was low, 78% and 58%, respectively. Positive and negative predicted values of CR were 78% and 58%, respectively and the number needed to diagnose was 2.777. Children in whom CR was performed (irrespective of the result) were at 22.9-fold higher risk of antibiotic therapy (95%CI: 14.1–37.1; $p < 0.01$), while those with a positive CR were only at 4.4-fold higher risk of antibiotic therapy (95%CI: 2.2–8.9; $p < 0.01$). Children with CR required a longer hospital stay than those without it (10 vs. 8 days, respectively; $p < 0.01$). The percentage of CR decreased from 78% in 2010 to 33% in 2017, with the lowest value of 11% in 2015. The additional cost of CR, which had no influence on treatment, would have been €381 had it been performed in each patient, which amounts to 1% of the total hospitalization cost. We conclude that CR is overused and in most cases it has no influence on the patient management. The recognition of practical meaning of CR is essential to avoid unnecessary radiation of children.

Keywords

Bronchiolitis · Chest radiography · Hospital stay · Respiratory syncytial virus · Respiratory tract infection

1 Introduction

Bronchiolitis is an acute disease of the lower respiratory tract in children under 2 years of age, which in most cases starts with signs and symptoms from the upper respiratory tract (e.g., rhinitis) and then progresses to involve the lower respiratory tract. Local inflammation, followed by necrosis of epithelial cells in small airways and increased mucus secretion cause a deterioration in the patient condition, including tachypnea, wheezing, or breathing difficulties. The clinical

A. Wrotek, M. Czajkowska, and T. Jackowska (✉)
Department of Pediatrics, Center of Postgraduate Medical Education, Warsaw, Poland

Department of Pediatrics, Bielanski Hospital, Warsaw, Poland
e-mail: tjackowska@cmkp.edu.pl

course of bronchiolitis is greatly variable, but many children require hospital treatment (Ralston et al. 2014). Especially, the youngest children are at higher risk of hospitalization, which is required in 25.9 per 1000 children aged under 1 month of age as opposed to 5.2 per 1000 children under 2 years of age (Hall et al. 2013). A vast majority (60–75%) of bronchiolitis are caused by the respiratory syncytial virus (RSV), but the proportion varies seasonally (AAP 2006). The recommended diagnostic tools are few. The diagnosis of bronchiolitis should be based on a physical examination and the patient's history. The main objective is to differentiate between bronchiolitis, which with a high probability is of a viral etiology, and other disorders, which may imitate bronchiolitis signs and symptoms (Ralston et al. 2014). Dawson et al. (1990) have shown no association between abnormalities in chest radiography (CR) and the clinical course of bronchiolitis and the authors suggest that CR should be reserved for the most severe cases, including intensive care unit transfer, a sudden deterioration of the condition, or a suspicion of co-morbidities.

The American Academy of Pediatrics (AAP 2006) in the first publication on bronchiolitis management does not recommend CR unless the above-mentioned severe complications arise. Likewise, the most recent AAP recommendations directly discourage from the use of routine chest X-rays in bronchiolitis (Ralston et al. 2014). In fact radiographic abnormalities may be seen in many patients with bronchiolitis. Farah et al. (2002) have reported pathological findings in CR in 17% of children with RSV infection, but their association with clinical outcomes is weak. Many a study emphasize that CR promotes the use of antibiotics due to the abnormalities revealed, with no major influence on the course of bronchiolitis (Schuh et al. 2007; Swingler et al. 1998). Akenroye et al. (2014) have shown that implementing the guidelines for bronchiolitis management in the emergency department decreases, inter alia, the number of CR by 23%, which also significantly contributes to cost reduction.

The general purpose of this study was to evaluate the use of CR in children with bronchiolitis.

We set out to estimate the frequency of CR use in RSV versus non-RSV infections during the 2010–2017 seasons, its correlation with the antibiotic use, the influence on treatment outcomes, and the economic impact.

2 Methods

Patients' files were retrospectively analyzed to estimate the frequency of CR and the influence of CR on disease management. Since the analysis was retrospective, clinical decisions undertaken at the time of hospitalization had not been influenced by the study protocol. There were 581 patients included in the analysis, hospitalized with RSV and a non-RSV infections between January 2010 and June 2017 (90 months). The group consisted of 347 boys (59.7%) and 234 girls (40.3%), aged from 8 days to 10 years, with a median of 3 months. The distribution of age groups was as follows: neonates – 68 (11.7%) patients, infants aged 1–3 months – 255 (43.9%), 4–6 months – 166 (28.6%), 7–12 months – 71 (12.2%), 13–24 months – 13 (2.2%), and over 24 months – 8 (1.4%). The final diagnosis of RSV bronchiolitis was in 390 (67.1%) and non-RSV bronchiolitis in 69 (11.9%) cases, followed by RSV pneumonia in 65 (11.2%) cases, and RSV bronchitis 57 (9.8%) cases.

Patients were divided into the following groups: CR group versus no-CR group, and CR positive result versus CR negative result. The groups were compared with respect to the clinical parameters and laboratory tests, which included: age, heart rate, breath rate, all assessed at admission, length of hospital stay, inflammatory markers such as white blood cell count (WBC), neutrophil and lymphocyte percentages, C-reactive protein (CRP), procalcitonin (PCT), and capillary blood gasometry (oxygen saturation, carbon dioxide pressure, and pH), and peripheral blood oxygen saturation and heart rate derived from pulse oximetry.

A final point in the assessment of CR usefulness was a change in the patient management, i.e., the implementation of antibiotic treatment or transfer to the pediatric intensive care unit or a tertiary reference center due to abnormalities found in the CR. If the CR findings influenced the management, CR results were marked as being true positive. If the CR was negative and had no influence on the disease management, CR results were marked as being true negative (CR correctly done but potentially unnecessary). The most commonly observed false positive and false negative CR results concerned the antibiotic implementation. The interpretation was the following: if there had been no abnormalities found in CR, but antibiotic treatment was initiated, the result was false negative. If, on the other side, there were any abnormal findings in CR but no antibiotic was given, the result was false positive. Patients requiring antibiotic treatment for other reasons than respiratory tract treatment, for instance, urinary tract infections, were not taken into account. The usefulness of CR was assessed on the basis of sensitivity, specificity, positive predictive value, negative predictive value, and the number needed to diagnose.

Since the Bielanski Hospital in Warsaw uses outsourcing radiological services, the mean cost of a single CR was calculated on the basis of a recent commercial rate. The cost of unnecessary CRs was estimated as a sum of false positive and false negative results multiplied by the cost of a single CR. Afterwards, a stimulation of costs was performed, with an assumption of CR use in each and every patient and a corresponding proportion of unnecessary costs calculated. The costs were then referred to the reimbursement that the hospital would obtain from the National Health Fund in Poland. The reimbursement policy is that the charge for hospitalization of each insured patient is returned directly to hospitals, but the sum is based on the final diagnosis, irrespective of the length of hospitalization or resources used.

Data distribution was checked with the Shapiro-Wilk test. Normally distributed data were expressed as means ±SD and non-normally distributed as medians with interquartile range (IQR). An unpaired t-test or the nonparametric Mann-Whitney U test were used as required. A multivariate logit model was used to show the relationship between CR (both its frequency and findings) and the aforementioned end points. The results were presented as odds ratios (OR) with a 95% confidence interval (95%CI). A p-value <0.05 defined statistically significant differences. A commercial statistical package of Statistica v13 (StatSoft; Tulsa, OK) was used for statistical analysis.

3   Results

CR was performed in 166 (28.6%) patients, including 19 CRs in the outpatient setting prior to hospitalization. The time analysis showed a strongly decreasing trend in conducting CR as 78% patients had CR in 2010, while the number dropped to 33% in 2017, with the lowest percentage of 11% noticed in 2015 (Fig. 1). Interestingly, frequency of CRs was much higher in patients with RSV bronchiolitis than in those with non-RSV bronchiolitis; 61% vs. 31%, respectively. There were 24 (14.5%) false positive and 24 (14.5%) false negative results. Overall, 29% of CRs failed to correlate with the subsequent treatment.

CR's prognostic sensitivity and specificity in guiding treatment was low; 78% and 58%, respectively. The CR positive and negative prognostic values also were rather low; 78% and 58%, respectively and the number needed to diagnose (NND) reached 2.777. The group of patients with CR was slightly older than that without it (median 3 vs. 2.4 months; p = 0.02). Patients who had CR performed, irrespective of result, had a 22.9-fold greater risk of antibiotic treatment (95%CI: 14.1–37.1; p < 0.01). In contrast, patients with positive CR had only a 4.4-fold greater risk of antibiotic treatment than those with negative CR (95%CI: 2.2–8.9, p < 0.01).

Moreover, patients with CR required a longer hospital stay (10 vs. 8 days; p < 0.01), and had a significantly higher WBC count (10.55 $*10^3$ cells/µL vs. 10.00 $*10^3$cells/µL; p = 0.019), higher neutrophil percentage (31.0% vs. 18.9%; p < 0.01), lower lymphocyte percentage (52.2% vs. 63.5%; p < 0.01), higher CRP level (2.62 mg/L vs. 0.90 mg/L; p < 0.01), and a higher procalcitonin concentration (0.13 ng/mL vs. 0.09 ng/mL; p < 0.01) than those without

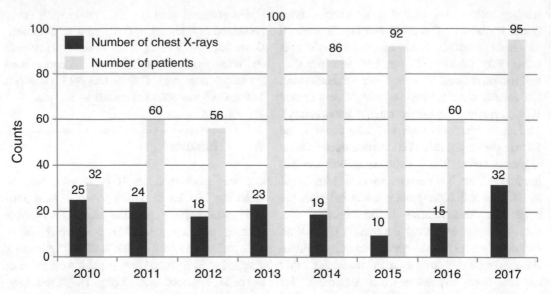

Fig. 1 Number of chest X-rays performed and the total number of patients in successive years

CR, respectively. Nevertheless, the above differences remained of no clinical significance. There were no differences in capillary blood-gas content, breath or heart rate, or peripheral capillary blood oxygen saturation (Table 1).

Comparing the patients with positive and negative CR results, the only difference was seen in the level of CRP (4.13 mg/L vs. 1.70 mg/L; $p < 0.01$); however, the higher value remained within the reference limit, and thus was of no clinical significance either. Patients with a positive CR also required a longer hospital stay (11 vs. 9 days; $p = 0.01$) (Table 2). Patients transferred to the pediatric intensive care unit did not differ in terms of CR frequency (6 out of the 10 patients had CR done, with two positive imaging results). However, CR results were of no clinical relevance either (Table 3).

The cost of all performed CRs reached €1,318, including €381 spent on CRs that did not influence the patient's management. If CR had to be performed in each patient with bronchiolitis or RSV infection, then the total cost would have reached €4,613, including €1,334 spent unnecessarily. The reimbursement for the whole group of patients (based on the clinical diagnosis) would have been approximately €440,134. The proportion of resources spent on CR, when done in each patient, would have been 1.05% of the whole sum, including at least 30% of the money spent needlessly.

4   Discussion

The most important question is of whether chest X-ray is at all needed in patients with bronchiolitis. Current guidelines, remain in line with the 2014 AAP recommendations that daunt a routine use of chest X-rays in bronchiolitis (Ralston et al. 2014). Yet statements to the contrary also exist. The Canadian Pediatric Society recommends the use of chest X-rays in case of uncertain diagnosis, lack of or slow improvement, as well as a suspicion of a bacterial pneumonia (Friedman et al. 2014). Special attention may apply the last condition above outlined, considering the risk of a bacterial suprainfection during RSV infection. Studies on theoretical models showed a higher risk of bacterial colonization in case of RSV (McGillivary et al. 2009, Hament et al. 2004), which has been lately confirmed in children hospitalized due to RSV infection (Suárez-Arrabal et al. 2015). The explanation for that may lie in the impact of RSV on the local

Table 1 Clinical and laboratory parameters in patients in whom chest radiography (CR) was and was not performed

Parameter	CR		No CR		
	Median	IQR	Median	IQR	p
Age (months)	3.0	1.5–5.9	2.4	1.5–4.0	<0.010
Length of stay (days)	10	8–13	8	7–10	<0.001
WBC ($*10^3$cells/µL)	10.55	8.60–15.15	10.00	8.20–12.50	0.020
Neu (%)	31.0	18.4–43.5	18.9	12.3–27.7	<0.001
Lym (%)	52.2	42.2–64.0	63.5	53.9–71.0	<0.001
CRP (mg/L)	2.62	0.65–13.53	0.90	0.26–3.47	<0.001
PCT (ng/dL)	0.13	0.09–0.34	0.09	0.07–0.12	<0.001
ScO_2	91.5	87.2–94.3	91.0	88.3–94.1	ns
PCO_2 (mmHg)	35.9	32.2–43.5	36.6	33.4–41.6	ns
pH	7.40	7.40–7.40	7.40	7.38–7.43	ns
BR (per min)	50	44–60	56	45–64	ns
HR (per min)	143	136–160	140	131–154	ns
SpO_2 (%)	95.0	93.0–97.0	96.0	94.0–97.0	ns

Data are medians and interquartile ranges (IQR); *CR* chest X-ray, *WBC* white blood cell count, *Neu* neutrophils, *Lym* lymphocytes, *CRP* C-reactive protein, *PCT* procalcitonin, ScO_2 capillary blood oxygen saturation, PCO_2 partial pressure of carbon dioxide and pH (gasometry), *BR* breathing rate, SpO_2 peripheral capillary oxygen saturation and *HR* heart rate (pulse oximetry, *ns* not significant

Table 2 Clinical and laboratory parameters in patients with positive and negative chest radiography (CR) results

Parameter	Positive CR result		Negative CR result		
	Median	IQR	Median	IQR	p
Age (months)	3.2	1.7–6.0	2.6	1.0–5.1	ns
Length of stay (days)	11	8–14	9	8–11	<0.01
WBC ($*10^3$cells/µL)	10.70	8.80–14.60	10.50	8.60–16.00	ns
Neu (%)	33.0	19.4–43.9	25.6	15.4–42.4	ns
Lym (%)	50.3	41.5–61.0	57.1	44.9–68.0	ns
CRP (mg/L)	4.13	1.15–17.67	1.70	0.41–6.84	<0.01
PCT (ng/dL)	0.13	0.09–0.40	0.14	0.10–0.20	ns
ScO_2	91.4	88.2–93.9	91.8	86.2–94.8	ns
pH	7.41	7.39–7.43	7.40	7.37–7.43	ns
PCO_2 (mmHg)	35.8	31.7–40.8	37.7	32.4–45.3	ns
BR (per min)	50	40–62	55	45–60	ns
HR (per min)	145	138–160	140	130–150	ns
SpO_2 (%)	95.0	93.0–98.0	95.0	93.0–96.0	ns

Data are medians and interquartile ranges (IQR); *WBC* white blood cell count, *Neu* neutrophils, *Lym* lymphocytes, *CRP* C-reactive protein, *PCT* procalcitonin, ScO_2 capillary blood oxygen saturation, PCO_2 partial pressure of carbon dioxide and *pH* (gasometry), *BR* breathing rate, SpO_2 peripheral capillary oxygen saturation and *HR* heart rate (pulse oximetry), *ns* not significant

microbiota, which modulates the host immune response (de Steenhuijsen Piters et al. 2016). A seasonal correlation between the frequency of an invasive pneumococcal disease and RSV infections is in line with this explanation (Weinberger et al. 2015; Techasaensiri et al. 2010; Ampofo et al. 2008; Jansen et al. 2008; Talbot et al. 2005). Further, facilitated adherence of *S. pneumoniae* to eukaryotic cells has been demonstrated in mice preinfected or coinfected with RSV and *S. pneumoniae*, which plausibly is due to a viral influence on the adhesins present on the S. *pneumoniae* membrane and cell wall (Hament et al. 2005).

Table 3 Characteristics of patients transferred to the pediatric intensive care unit; all diagnoses of bronchiolitis

Year	Patient age	Chest X-ray	Elevated CRP	Antibiotic treatment
2010	39 months	no	no	–
2010	3 months	**Positive**	80.0 mg/L	Cefuroxime
2012	16 days	no	no	–
2017	27 days	Negative	no	Cefotaxime
2017	17 days	Negative	no	–
2011	1 month	Negative	no	–
2014	1 month	no	no	–
2016	3 months	**Positive**	no	Cefotaxime
2013	13 days	Negative	no	–
2013	1 month	no	no	–

On the basis of the present finding we cannot confirm or exclude any bacterial suprainfection. Differences we noticed between patients with positive and negative results of CR were few and of no clinical meaning. The role of additional laboratory investigations, such as CRP or PCT, in differentiating bacterial from viral pneumonia is widely discussed, but the prevailing consensus is that acute phase reactants cannot reliably set the etiology of pneumonia (Bradley et al. 2011; Harris et al. 2011). At any rate, we did not notice any case of a serious bacterial infection in this study. The only appreciable difference between patient who had and did not have CR done concerned the length of hospitalization, with the former having a longer stay than the latter (10 vs. 8 days, respectively). Likewise, patients with a positive finding in CR images were hospitalized longer than those with a negative finding (11 days vs. 9 days, respectively). These differences in the length of hospitalization might, in fact, be spurious, as CR could be done due to the patient's worse condition, which enforced a longer hospitalization, rather than due to the mere fact of doing CR, or its positive or negative finding. Interestingly, almost a two-fold greater frequency of CR was observed in the RSV group, as opposed to the non-RSV group of patients in this study (61% vs. 31%, respectively). That is in line with the notion that a doing of CR might be related with the patient condition as RSV is recognized as a frequent cause of severe lower respiratory tract infections, and of longer hospital treatment (Ramagopal et al. 2016; Stollar et al. 2014).

A correlation between the chest X-ray imaging and the antibiotic use reported in a randomized controlled trial by Swingler et al. (1998) is widely cited concerning the children with bronchiolitis (Friedman et al. 2014; Ralston et al. 2014). However, the setting of that study was outpatient, and the patients were diagnosed with pneumonia. In the present study, CR was done to bring about changes in the patient's management, and then the antibiotic use was taken as a true positive result when it associated with the CR result. We do not support the use of antibiotics in each case of abnormalities found in CR, although that might be justified in some cases. We aimed to show how frequently CR did not influence the patient's management and may thus be interpreted as overused.

The presence or suspicion of co-morbidities is another occurrence of CR examination in young children. For instance, children with a hemodynamically unstable heart disorder are at a higher risk of severe RSV course. The frequency of a congenital heart disease is 8 per 1,000 live births, and approximately one half of those children presents a hemodynamic impairment (Baraldi et al. 2014). Some of those children remain asymptomatic for a longer period of time, and the first symptoms may appear during RSV infection, which typically develops in the first year of life. A prospective study by Schuh et al. (2007) has shown that only in 2 out of the 265 children a chest X-ray reveals abnormalities other than those related to bronchiolitis, such as cardiomegaly and lobar pneumonia. We found no instances of congenital heart disease in the present study.

Therefore, we submit that the use of CR imaging as a diagnosing tool for unraveling the underlying congenital heart disease in children coming down with bronchiolitis is largely unjustified. To this end, it also is worthwhile to recall that children have a faster cell division and are definitely more sensitive to radiation (Thomas et al. 2006; Huda 2004). Thus, every effort ought to be performed to decrease the dose and frequency of irradiation, best by its avoidance if not absolutely required.

In conclusion, chest radiography is generally overused in children with bronchiolitis. On the positive side, a strongly increased use of antibiotics in patients in whom chest radiography is done while it is much less increased in case of positive radiographic findings, a higher percentage of chest X-rays in RSV-bronchiolitis usually having a more severe course than that in non-RSV bronchiolitis, and a longer hospital stay of patients in whom chest radiography is done, all lead to a consistent impression that chest radiography is used with restrain and mostly in severely ill patients. Therefore, recognizing the practical bearing of chest radiography, it is essential to avoid unnecessary irradiation. Chest radiographs in children with bronchiolitis should not be a routine clinical practice. If X-rays were performed sparingly, then their use in specific cases would be justified.

Acknowledgements Supported by CMKP grant 501-1-020-19-19.

Conflicts of Interest The authors declare no conflict of interest in relation to this article.

Ethical Approval All procedures performed in studies involving human participants were in accordance with the ethical standards of the institutional and/or national research committee and with the 1964 Helsinki declaration and its later amendments or comparable ethical standards. The study protocol was approved by an institutional Ethics Committee.

Informed Consent The study had a retrospective character of reviewing the patients' hospital files. Since there was no direct contact with the study participants, the requirement of obtaining informed consent from each individual was waived.

References

AAP (2006) American Academy of Pediatrics Subcommittee on Diagnosis and Management of Bronchiolitis. Diagnosis and management of bronchiolitis. Pediatrics 118:1774–1793

Akenroye AT, Baskin MN, Samnaliev M, Stack AM (2014) Impact of a bronchiolitis guideline on ED resource use and cost: a segmented time-series analysis. Pediatrics 133:e227–e234

Ampofo K, Bender J, Sheng X, Korgenski K, Daly J, Pavia AT, Byington CL (2008) Seasonal invasive pneumococcal disease in children: role of preceding respiratory viral infection. Pediatrics 122:229–237

Baraldi E, Lanari M, Manzoni P, Rossi GA, Vandini S, Rimini A, Romagnoli C, Colonna P, Biondi A, Biban P, Chiamenti G, Bernardini R, Picca M, Cappa M, Magazzù G, Catassi C, Urbino AF, Memo L, Donzelli G, Minetti C, Paravati F, Di Mauro G, Festini F, Esposito S, Corsello G (2014) Inter-society consensus document on treatment and prevention of bronchiolitis in newborns and infants. Ital J Pediatr 40:65

Bradley JS, Byington CL, Shah SS, Alverson B, Carter ER, Harrison C, Kaplan SL, Mace SE, McCracken GH Jr, Moore MR, St Peter SD, Stockwell JA, Swanson JT, Pediatric Infectious Diseases Society and the Infectious Diseases Society of America (2011) The management of community-acquired pneumonia in infants and children older than 3 months of age: clinical practice guidelines by the Pediatric Infectious Diseases Society and the Infectious Diseases Society of America. Clin Infect Dis 53:e25–e76

Dawson KP, Long A, Kennedy J, Mogridge N (1990) The chest radiograph in acute bronchiolitis. J Paediatr Child Health 26:209–211

de Steenhuijsen Piters WA, Heinonen S, Hasrat R, Bunsow E, Smith B, Suárez-Arrabal MC, Chaussabel D, Cohen DM, Sanders EA, Ramilo O, Bogaert D, Mejias A (2016) Nasopharyngeal microbiota, host transcriptome, and disease severity in children with respiratory syncytial virus infection. Am J Respir Crit Care Med 194:1104–1115

Farah MM, Padgett LB, McLario DJ, Sullivan KM, Simon HK (2002) First-time wheezing in infants during respiratory syncytial virus season: chest radiograph findings. Pediatr Emerg Care 18:333–336

Friedman JN, Rieder MJ, Walton JM, Canadian Paediatric Society, Acute Care Committee, Drug Therapy and Hazardous Substances Committee (2014) Bronchiolitis: recommendations for diagnosis, monitoring and management of children one to 24 months of age. Paediatr Child Health 19:485–498

Hall CB, Weinberg GA, Blumkin AK, Edwards KM, Staat MA, Schultz AF, Poehling KA, Szilagyi PG, Griffin MR, Williams JV, Zhu Y, Grijalva CG, Prill MM, Iwane MK (2013) Respiratory syncytial virus-associated hospitalizations among children less than 24 months of age. Pediatrics 132:e341–e348

Hament JM, Aerts PC, Fleer A, Van Dijk H, Harmsen T, Kimpen JL, Wolfs TF (2004) Enhanced adherence of *Streptococcus pneumoniae* to human epithelial cells infected with respiratory syncytial virus. Pediatr Res 55:972–978

Hament JM, Aerts PC, Fleer A, van Dijk H, Harmsen T, Kimpen JL, Wolfs TF (2005) Direct binding of respiratory syncytial virus to pneumococci: a phenomenon that enhances both pneumococcal adherence to human epithelial cells and pneumococcal invasiveness in a murine model. Pediatr Res 58:1198–1203

Harris M, Clark J, Coote N, Fletcher P, Harnden A, McKean M, Thomson A, British Thoracic Society Standards of Care Committee (2011) British Thoracic Society guidelines for the management of community acquired pneumonia in children: update 2011. Thorax 66:ii1–ii23

Huda W (2004) Assessment of the problem: pediatric doses in screen–film and digital radiography. Pediatr Radiol 34(Suppl 3):S173–S182

Jansen AG, Sanders EA, van der Ende A, van Loon AM, Hoes AW, Hak E (2008) Invasive pneumococcal and meningococcal disease: association with influenza virus and respiratory syncytial virus activity? Epidemiol Infect 136:1448–1454

McGillivary G, Mason KM, Jurcisek JA, Peeples ME, Bakaletz LO (2009) Respiratory syncytial virus-induced dysregulation of expression of a mucosal beta-defensin augments colonization of the upper airway by non-typeable *Haemophilus influenza*. Cell Microbiol 11:1399–1408

Ralston SL, Lieberthal AS, Meissner HC, Alverson BK, Baley JE, Gadomski AM, Johnson DW, Light MJ, Maraqa NF, Mendonca EA, Phelan KJ, Zorc JJ, Stanko-Lopp D, Brown MA, Nathanson I, Rosenblum E, Sayles S 3rd, Hernandez-Cancio S, American Academy of Pediatrics (2014) Clinical practice guideline: the diagnosis, management, and prevention of bronchiolitis. Pediatrics 134:e1474–e1502

Ramagopal G, Brow E, Mannu A, Vasudevan J, Umadevi L (2016) Demographic, clinical and hematological profile of children with bronchiolitis: a comparative study between respiratory syncytial virus (RSV) and (Non RSV) groups. J Clin Diagn Res 10:SC05–SC08

Schuh C, Lalani A, Allen U, Manson D, Babyn P, Stephens D, MacPhee S, Mokanski M, Khaikin S, Dick P (2007) Evaluation of the utility of radiography in acute bronchiolitis. J Pediatr 150:429–433

Stollar F, Alcoba G, Gervaix A, Argiroffo CB (2014) Virologic testing in bronchiolitis: does it change management decisions and predict outcomes? Eur J Pediatr 173:1429–1435

Suárez-Arrabal MC, Mella C, Lopez SM, Brown NV, Hall MW, Hammond S, Shiels W, Groner J, Marcon M, Ramilo O, Mejias A (2015) Nasopharyngeal bacterial burden and antibiotics: influence on inflammatory markers and disease severity in infants with respiratory syncytial virus bronchiolitis. J Infect 71:458–469

Swingler G, Hussey G, Zwarenstetin M (1998) Randomised controlled trial of clinical outcome after chest radiograph in ambulatory acute lower respiratory tract infection in children. Lancet 345:404–408

Talbot TR, Poehling KA, Hartert TV, Arbogast PG, Halasa NB, Edwards KM, Schaffner W, Craig AS, Griffin MR (2005) Seasonality of invasive pneumococcal disease: temporal relation to documented influenza and respiratory syncytial virus circulation. Am J Med 118:285–291

Techasaensiri B, Techasaensiri C, Mejías A, McCracken GH Jr, Ramilo O (2010) Viral coinfections in children with invasive pneumococcal disease. Pediatr Infect Dis J 29:519–523

Thomas KE, Parnell-Parmley JE, Haidar S, Moineddin R, Charkot E, BenDavid G, Krajewski C (2006) Assessment of radiation dose awareness among pediatricians. Pediatr Radiol 36:823–832

Weinberger DM, Klugman KP, Steiner CA, Simonsen L, Viboud C (2015) Association between respiratory syncytial virus activity and pneumococcal disease in infants: a time series analysis of US hospitalization data. PLoS Med 12:e1001776

Advs Exp. Medicine, Biology - Neuroscience and Respiration (2019) 44: 63–68
https://doi.org/10.1007/5584_2019_424
© Springer Nature Switzerland AG 2019
Published online: 28 July 2019

Evaluation of the 2017/18 Influenza Epidemic Season in Poland Based on the *SENTINEL* Surveillance System

K. Łuniewska, K. Szymański, E. Hallmann-Szelińska, D. Kowalczyk, R. Sałamatin, A. Masny, and L. B. Brydak

Abstract

The *SENTINEL* influenza surveillance system is an important tool for monitoring influenza in Poland. Data from this system are necessary to determine the dynamics of seasonal infections and to announce the epidemic by the country level. For the 2017/18 epidemic season, the dominance of influenza type B was recorded and the highest percentage of infections was recorded in the age group 45–64 years. Among the subtypes of influenza type A, A/H1N1/pdm09 was the predominated subtype. Most cases were reported in the age group of 26–44 and 0–4 years. The influenza virus frequently undergoes modifications. Therefore, it is necessary to constantly monitor the emerging strains around the world.

Keywords

Epidemic season · Influenza · Respiratory tract · SENTINEL system · Surveillance · Virology

K. Łuniewska (✉), K. Szymański, E. Hallmann-Szelińska, D. Kowalczyk, A. Masny, and L. B. Brydak
Department of Influenza Research, National Influenza Center, National Institute of Public Health- National Institute of Hygiene, Warsaw, Poland
e-mail: kluniewska@pzh.gov.pl

R. Sałamatin
Department of General Biology and Parasitology, Warsaw Medical University, Warsaw, Poland

1 Introduction

The *SENTINEL* Influenza Surveillance System allows monitoring of the course and dynamics of epidemic seasons in Poland. Virological data tallied by 16 Voivodship Sanitary and Epidemiological Stations of the country allow to identify the dominant virus strain and to provide information on the age groups that are most vulnerable to the virus (Cieślak et al. 2017). Monitoring of the influenza virus activity in the inter-pandemic period also is a part of the national surveillance plan in Poland. The information about the emergence of a new pandemic strain can be obtained immediately, and thus the anti-influenza measures implemented at the national level (Bednarska et al. 2016).

In 2009, the EU Commission has issued recommendations in which it is assumed that the influenza vaccination rate in high-risk groups should reach 75% in all European countries up to the 2014/15 epidemic season. The risks groups included adults above 65 years of age and all persons over 6 months of age with chronic diseases (ECDC 2017). However, this goal has not been achieved in most of the countries (Weimbergen 2018). The vaccination rate in the 2017/18 epidemic season in the population of Poland amounted to 3.4%. Such a dismal vaccination rate places the country in the last place in Europe (Brydak 2019). Further, seasonal epidemics occur in Poland every year, causing a large number of deaths and post-influenza morbidity, incurring huge socioeconomic costs (Brydak 2018). The epidemics cannot be properly countered without a proper population vaccination rate, the main preventive measure against influenza. This study seeks to define the dynamics of influenza infections and the prevalence of dominant viral types in Poland during the 2017/18 epidemic season, based on the data tallied by the *SENTINEL* influenza surveillance program.

2 Methods

The study material included nasal and pharyngeal swab samples analyzed in 16 Voivodship Sanitary and Epidemiological Stations and in the Department of Influenza Research, National Influenza Center in National Institute of Public Health – National Institute of Hygiene. Data were analyzed and reported using the *SENTINEL* Influenza Surveillance System. A total of 1,585 patients participated in the study. They were divided into 7 age groups: 0–4, 5–9, 10–14, 15–25, 26–44, 45–64, and 65+ years.

Specimens collected from patients were analyzed for the identification of influenza type A and B viruses by PCR techniques. From a 200 μl clinical sample suspended in physiological saline, 50 μl of viral RNA resuspended in RNase-free water was obtained. For the assay, a Maxwell 16 Viral Total Nucleic Acid Purification Kit was used (Promega Corporation; Madison, WI) according to the manufacturer's instructions. The analytes were further analyzed to determine viral subtypes using a Light Thermocycler 2.0 System (Roche Diagnostics; Rotkreuz, Switzerland). The primers and probes were obtained from the International Reagent Resource run by the Centers for Disease Control and Prevention (CDC). The reaction was carried out in accordance with the manufacturer's instructions. To obtain cDNA, RNA was subjected to reverse transcription (at 50 °C for 30 min). Then, cDNA was subjected to the initiating process (1 cycle of 95 °C for 2 min) followed by 45 cycles of amplification: denaturation at 95 °C for 15 s, annealing at 55 °C for 10 s, and elongation at 72 °C for 20 s. Positive control constituted viral RNA obtained from the vaccine strains for the current epidemic season (A/Michigan/45/2015 (H1N1)pdm09, A/HongKong/4801/2014 (H3N2), B/Brisbane/60/2008), and negative control constituted RNase-free water.

Using the reverse transcription polymerase chain reaction, the presence of 15 respiratory viruses was confirmed in the samples. The viruses consisted of Influenza A virus, Influenza B virus, Human respiratory syncytial virus A and B, Human adenovirus, Human metapneumovirus, Human coronavirus 229E/NL63, Human coronavirus OC43, Human parainfluenza 1, 2, 3, and 4, Human rhinovirus A/B/C, Human enterovirus, and Human bocavirus 1/2/3/4. For the assay, RV15 OneStep ACE Detection Kit (Seeplex; Seoul, South Korea) was used according to the manufacturer's instructions. When the reaction was completed, the product was separated on a 2% agarose gel by electrophoresis.

The probability of occurrence of influenza infection was estimated using a logistic regression model. Odds ratios were calculated for all age groups with reference to the 65+ group. The statistical analysis was performed using commercial IBM SPSS statistical software (Armonk, NY).

3 Results

In the 2017/2018 epidemic season, a total of 1,585 samples, one from each individual, were collected through the *SENTINEL* Influenza Surveillance System. The overall number of analyzed samples remained at a low level until Week 1 of 2018, not exceeding 24 per week. Later on, there was a rapid increase in the number of tested samples, which peaked in Week 8 reaching 196 samples, except for Week 7 when this number retreat to 128 samples. The situation was different when we analyzed the number of positive samples in a given week. The first positive samples were recorded at Week 48 of 2017 (15.8% of weekly samples). A steady increase in influenza cases was reported from Week 1 of 2018 (23.5% positive), exceeding 47.0% of the samples tested in Week 11. The highest percentage of positive samples was recorded at Week 10 of 2018, which amounted to 65.6%. There was a decrease in the percentage of positive findings in the following weeks until Week 16, when there was a drop to zero noted (Fig. 1).

Comparing the recent epidemic seasons starting from 2014/15 up to 2017/18, an increase in the number of confirmed cases of influenza from 221 to 708 cases, respectively, is observed. An increase also is observed in the percentage terms, from 33.8% to 44.7%, respectively, although this increase has plateaued in the last season as the greatest, amounting to 46.6%, was observed in before last 2016/17 epidemic season (Table 1).

In then 2017/18 epidemic season, a total of 708 influenza infections and 28 influenza-like viral infections were confirmed. Of the 708 influenza infections, 204 were due to influenza A virus and 504 due to influenza B virus. Influenza A infections were dominated by the subtype A/H1N1/pdm09 (38 cases). Influenza-like infections viruses were dominated by respiratory syncytial virus (16 cases).

The findings of this study were reanalyzed in 7 successive age groups. The highest percentage of influenza B infections was found in the age group of 45–64 (21.6%) years, while that of influenza A in the group of 26–44 (7.6%) years. The dominance of influenza A over B was found only in the youngest age group of 0–4 years. There was a distinct dominance of influenza B virus in the remaining age groups (Fig. 2).

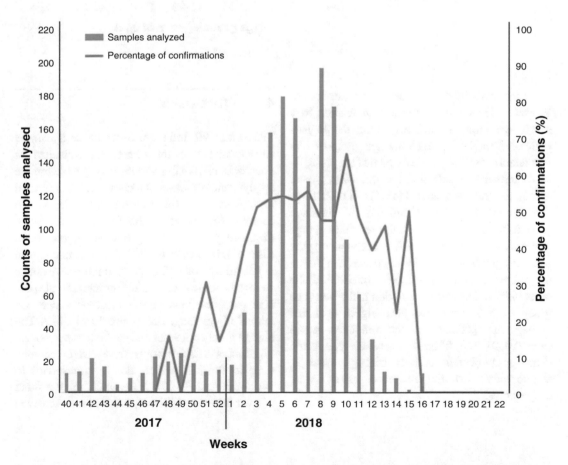

Fig. 1 Dynamics of the 2017/18 influenza epidemic season based on the SENTINEL surveillance system in Poland

Table 1 Influenza infections in recent epidemic seasons

Epidemic season	2014/15	2015/16	2016/17	2017/18
Number of samples analyzed	653	1,625	1,283	1,585
Number of confirmed cases	221	603	598	708
Percentage of confirmations	33.8%	37.1%	46.6%	44.7%

Fig. 2 Percentage of confirmed influenza infections in the 2017/18 epidemic season in successive age groups of patients

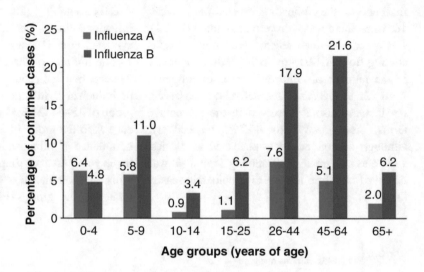

The probability of occurrence of influenza A or B virus infection was estimated in relation to a given age group of patients, using the logistic regression model in which the age group of 65+ was taken as reference. We found that the probability of influenza A infection was three-fold greater among the patients aged 0–4 (p = 0.001) and it was 50% lower than that of influenza B infection (p = 0.028), compared with the reference group. The probability of influenza A infection among the patients aged 5–9 was 2.5-fold greater than that in the reference group (p = 0.005). In addition, the probability of influenza B infection in the same age group of 5–9 years was 58% higher than that among the patients in the reference group (p = 0.049). The differences between type A and B influenza infections in the remaining age groups, as compared to the reference group, were insignificant.

4   Discussion

The *SENTINEL* Influenza Surveillance System is an essential tool for monitoring influenza in Poland. Data from this system are key to determine the dynamics of season infections and to announce the epidemic by the relevant authorities in the country (Cieślak et al. 2018; Lee et al. 2018). Comparing the increasing number of confirmed cases of influenza in the 2017/18 epidemic season in Poland with other European countries, we notice almost the same situation. In Germany, single positive samples have been recorded since the season's beginning, i.e., Week 40 of 2017. The highest number of cases and the highest percentage of positive samples (over 60%) were recorded there in Week 9, which closely corresponds to the Polish data. In Czech Republic, the threshold of over 60% of positive specimens was achieved already in Week 5 and lasted until Week 9. Further, the dominance of influenza type B has been reported there, similarly to the epidemiological situation in Poland. Concerning influenza type A, subtype A/H1N1/pdm09 has also dominated (FluNews Europe 2018). This trend has been different from that in the 2016/17 season, when the subtype A/H3N2/dominated, while the A/H1N1/pdm09 was not recorded (Cieślak et al. 2018). The first positive specimens in Poland were recorded in Week 48, which distinguishes the current season from the previous one. The 2016/17 season started earlier, because the threshold of 10% of positive tests has already been exceeded from Week 46 onward (Adlhoch et al. 2018).

In the 2017/18 epidemic season, the dominance of subtype A/H1N1/pdm09 among the influenza A subtypes was noticed. Most cases were reported among the patients of 26–44 and 0–4 years of age. A similar situation took place in the course of the 2009 pandemic when the subtype A/H1N1/pdm09 had infected mainly children and young adults, while it was responsible for just 10% of hospitalization of the elderly (Verma et al. 2012). In the previous 2016/17 influenza season, A/H3N2/ dominated in terms of the percentage of influenza cases; a subtype that is known for its propensity to infect the elderly (Pebody et al. 2017).

Over the last four influenza epidemic seasons, we have observed an increase in the percentage of confirmed influenza infections reported in the *SENTINEL* Surveillance System. That shows a steady improvement in the surveillance system and in the knowledge and ability to recognize and verify the infection and its type. In the last two influenza seasons of 2016/17 and 2017/18, the percentage of confirmed cases ranged around 45%, which means that almost every second patient was properly diagnosed by general practitioners.

The present report demonstrates the dynamics of the 2017/18 influenza season in Poland according to the data from Sentinel Surveillance System. The virological data are valuable in that they serve to accurately determine the antigenic composition of vaccine for the next epidemic season.

Acknowledgments We thank physicians and employees of VSES who participated in the *SENTINEL* program for their input in the influenza surveillance in Poland. Supported by NIPH-NIH theme 3/EM.

Conflicts of Interest The authors declare no conflicts of interests in relation to this article.

Ethical Approval All procedures performed in studies involving human participants were in accordance with the ethical standards of the institutional and/or national research committee and with the 1964 Helsinki declaration and its later amendments or comparable ethical standards. The study was approved by the Ethics Committee of the NIPH-NIH in Warsaw, Poland.

Informed Consent Informed consent for taking nasopharyngeal swab specimens was obtained from all patients included in the study or their legal quardians at the time of sampling.

References

Adlhoch C, Snacken R, Melidou A, Ionescu S, Penttinen P, The European Influenza Surveillance Network (2018) Dominant influenza A(H3N2) and B/Yamagata virus circulation in EU/EEA, 2016/17 and 2017/18 seasons, respectively. Eur Secur 23 (13):18–00146

Bednarska K, Hallmann-Szelińska E, Kondratiuk K, Brydak LB (2016) Influenza surveillance. Postepy Hig Med Dosw 70:313–318

Brydak LB (2018) Health and economic impacts of influenza infections in the context of public health in Poland. In: Nowakowska E (ed) Pharmacoeconomics in the management of health care resources. Wolters Kluwer Poland. (Article in Polish)

Brydak LB (2019) Flu – prevention and treatment in children and adolescents. Pediatr Med Stand 2:16

Cieślak K, Kowalczyk D, Szymański K, Brydak LB (2017) The Sentinel system as the main influenza surveillance tool. Adv Exp Med Biol 980:37–43

Cieślak K, Szymański K, Kowalczyk D, Hallmann-Szelińska E, Brydak LB (2018) Virological situation in Poland in the 2016/2017 epidemic season based on Sentinel data. Adv Exp Med Biol 1108:63–67

ECDC (2017) European centre for disease prevention and control. Seasonal influenza vaccination in Europe. Vaccination recommendations and coverage rates in the EU member states for eight influenza seasons: 2007–2008 to 2014–2015. https://ecdc.europa.eu/sites/portal/files/documents/influenzavaccination-2007%E2%80%932008-to-2014%E2%80%932015.pdf. Accessed on 26 June 2018

FluNews Europe (2018). https://flunewseurope.org/. Accessed on 26 June 2019

Lee EC, Arab A, Goldlust SM, Viboud C, Grenfell BT, Bansal S (2018) Deploying digital health data to optimize influenza surveillance at national and local scales. PLoS Comput Biol 14(3):e1006020

Pebody R, Warburton F, Ellis J, Andrews N, Potts A, Cottrell S, Reynolds A, Gunson R, Thompson C, Galiano M, Robertson C, Gallagher N, Sinnathamby M, Yonova I, Correa A, Moore C, Sartaj M, de Lusignan S, McMenamin J, Zambon M (2017) End-of-season influenza vaccine effectiveness in adults and children, United Kingdom, 2016/17. Eur Secur 22(44):17–00306

Verma N, Dimitrova M, Carter DM, Crevar CJ, Ross TM, Golding H, Khurana S (2012) Influenza virus H1N1pdm09 infections in the young and old: evidence of greater antibody diversity and affinity for the hemagglutinin globular head domain (HA1 domain) in the elderly than in young adults and children. J Virol 86 (10):5515–5522

Weimbergen B (2018) Vaccines for elderly: current use and future challenges. Immun Ageing 15:3

Advs Exp. Medicine, Biology - Neuroscience and Respiration (2019) 44: 69–73
https://doi.org/10.1007/5584_2019_443
© Springer Nature Switzerland AG 2019
Published online: 22 October 2019

Occurrence of Influenza Hemagglutinin Antibodies in the Polish Population during the Epidemic Season 2017/18

Ewelina Hallmann-Szelińska, K. Szymański, K. Łuniewska, A. Masny, D. Kowalczyk, R. Sałamatin, and L. B. Brydak

Abstract

This study seeks to define the level of antihemagglutinin antibodies, using the hemagglutination inhibition assay (HAI), in the serum of patients, stratified into seven age groups, in Poland during the influenza epidemic season of 2017/18. A quadrivalent influenza vaccine has been introduced in Poland as of this epidemic season, making it possible for the first time to conduct the analysis for four antigens: A/Michigan/45/2015 (H1N1) pdm09, A/Hong Kong/4801/2014 (H3N2), B/Brisbane/60/2008 – Victoria lineage, and B/Phuket/3073/2013 – Yamagata lineage. We found that the level of individual antihemagglutinin antibodies was different among the seven age groups studied; with the highest in patients of 5–9 years and 10–14 years of age. Interestingly, the protection factor, defined as the percentage of people with the level of antihemagglutinin antibodies of at least 1:40 after vaccination or due to a previous infection, was the highest for the antigen A/Hong Kong/4801/2014 (H3N2) in the same age groups (74% and 75%, respectively). Taking into account the dismal 3.6% of the vaccinated population in Poland, these findings point toward the sustained presence of an immune system response in patients after a prior influenza virus infection.

Keywords

Epidemic season · Hemagglutinin antibodies · Immune system · Influenza · Protection factor · Respiratory infection

E. Hallmann-Szelińska (✉), K. Szymański,
K. Łuniewska, A. Masny, D. Kowalczyk,
and L. B. Brydak
Department of Influenza Research – National Influenza Center, National Institute of Public Health – National Institute of Hygiene, Warsaw, Poland
e-mail: ehallmann@pzh.gov.pl

R. Sałamatin
Department of General Biology and Parasitology, Warsaw Medical University, Warsaw, Poland

1   Introduction

Vaccination is by far the only effective method of prevention of influenza. According to the recommendations of the Advisory Committee on Immunization Practices of the World Health Organization (ACIP 2016), high-risk group of individuals should be vaccinated before the influenza virus starts its active circulation in the population, which in Poland comes up in October. The purpose of seasonal vaccination against influenza is to produce a protective concentration of antibodies. Antibodies detected in unvaccinated individuals indicate a history of past infection (Brydak 2008). In the 2017/18 epidemic season, a peak incidence of influenza, taking the number of positive clinical samples into account, occurred in Week 8 (February 19–25) of 2018. Overall, there were 5,182,291 cases of influenza and influenza-like illnesses recorded in the season, with the incidence of 13,637 per 100,000 inhabitants. Influenza B virus, the Yamagata lineage, was the predominated strain in the 2017/18 season in Poland. The influenza-like virus was dominated by a respiratory syncytial virus (RSV). Starting as of the 2017/18 season, a four-component vaccine, consisting two antigens of influenza type A virus and two antigens of influenza type B virus, has become available. Unfortunately, the percentage of vaccinated population gets lower in Poland season by season. In the 2017/18 season, only 3.6% of the population was vaccinated (NIPH-NIH 2018). Therefore, this study seeks to define the level of specific antibodies against the influenza virus hemagglutinin in the serum of patients, stratified by successive age groups, in Poland during the 2017/18 influenza season.

2   Methods

Serum samples of individuals belonging to seven successive age groups (0–4, 5–9, 10–14, 15–25, 26–44, and 45–64 years of age) were collected by the employees of the Voivodship Sanitary Epidemiological Stations across the country and were then sent to the Department of Influenza Research, National Influenza Center of the National Institute of Public Health – National Institute of Hygiene (NIPH-NIH) in Warsaw, Poland. A total of 1,050 randomly selected sera, 150 samples in each age group, were stored at −80 °C until tested.

The antigens listed in Table 1, according to the WHO   recommendations   for   the   2017/18

epidemic season (WHO 2017), were prepared in-house for testing. Hemagglutination inhibition test was used with 8 hemagglutination units (WHO 2011; Pedersen 2014). Prior to analysis, the sera were inactivated using the *Vibrio cholerae* enzyme and were serially diluted from 1:10 to 1:1280, according to the method of Brydak et al. (2003).

Data were presented as geometric mean titers (GMT) for antihemagglutinin antibodies.

3   Results and Discussion

The GMTs for antihemagglutinin antibodies, calculated for the successive age groups of patients in the 2017/18 epidemic season in Poland are shown in Fig. 1. The GMT levels of all antihemagglutinin antibodies were similar in children aged 0–4 years, ranging from about 47 to 52. These values were the highest for hemagglutinin H3 in children aged 5–9 and 10–14 years; 122.0 and 136.8, respectively. Among adults, the level of the antibodies for hemagglutinin H3 was below 40, with the lowest of 14.9 in people aged 26–44 years. Overall, GMTs for antihemagglutinin antibodies were lower in adults than those in children, with the most distinct downward difference noticed in case of hemagglutinin H3.

The protection factor is defined as the percentage of people with a protective level of antihemagglutinin antibodies of at least 1:40 after vaccination or due to previous infection (Brydak et al. 2003). It should be noted that for people 60+ years of age, the protection factor is expected to be ≥60%, whereas for people aged 18–60 years, it should be ≥70% (Brydak 2008). In the 2017/18 epidemic season, the highest protection levels were recorded for hemagglutinin

Table 1 Influenza virus antigens used for the hemagglutination inhibition test (HAI)

Strains of influenza virus				
Epidemic season 2017/18	A/H1N1/	A/H3N2/	B Victoria lineage	B Yamagata lineage
	A/Michigan/45/2015 (H1N1) pdm09-like virus	A/Hong Kong/4801/2014 (H3N2)-like virus	B/Brisbane/60/2008	B/Phuket/3073/2013

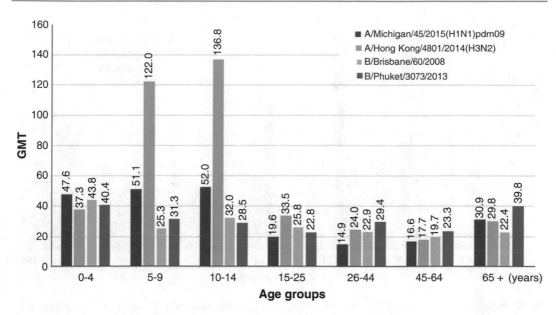

Fig. 1 Geometric mean titers (GMT) of antihemagglutinin antibodies in successive age groups in the influenza epidemic season 2017/18 in Poland

strain A/H3N2/−A/Hong Kong/4801/2014, which was 74% in children 5–9 years and 75% in 10–14 years old. In the remaining age groups, the protection factor was lower, ranging from about 21% to 47% (Fig. 2). For hemagglutinin H1 strain A/Michigan/45/2015 (H1N1) pdm09, the highest protection level was recorded in children 5–9 years (31%) and 10–14 years old (38%) and the lowest in people 65+ years old (29%). In the remaining four age groups, the protection factor was below 12%. For hemagglutinin B strain B/Brisbane/60/2008 like virus of the Victoria line, the highest protection factor was in children 0–4 years (39%) old and the lowest in people 45–64 years old (13%). In the remaining age groups, the protection factor ranged from 19% (26–44 years) to 37% (10–14 years). For hemagglutinin B strain B/Phuket/3073/2013 like virus of the Yamagata line, the highest levels of protection factor, in a range of 40–47% were in the age groups of 0–4, 16–44, and 65+ years of age and the lowest in the age group of 45–64 years (25%).

A four-component influenza vaccine introduced in the 2017/18 epidemic season had a modified antigenic content, compared to past immunizations, in particular the A/California/7/ 2009 (H1N1) pdm09 antigen, a vaccine component since 2010/11, had been removed. The vaccine provided different levels of GMT for antihemagglutinin antibodies and consequently different protection factors when compared with the preceding 2016/17 season. In the 2016/17 season, the highest GMT value for type B hemagglutinin of the strain B/Brisbane/60/2008 had been recorded in the age group of 10–14 years (51.6), whereas it had been the highest (136.8) in the same age group but for hemagglutinin H3 in the current 2017/18 season. In the main, higher values of GMT had been recorded in the previous season when compared with the current one, which was particularly noticed in all adult age groups (Hallmann-Szelińska et al. 2018). Concerning the protection factor, it was the highest for subtype A/H3N2/ in children of 10–14 years (75%) and the lowest in adults of 45–64 years of age (4.0%) in the 2017/18, compared with the highest of 60% also in children of 10–14 years and the lowest of about 15% in children of 5–9 years and adults of 26–44 years and 65+ years of age in the 2016/17 season (Kowalczyk et al. 2019). On the positive side, considering that children, particularly of young

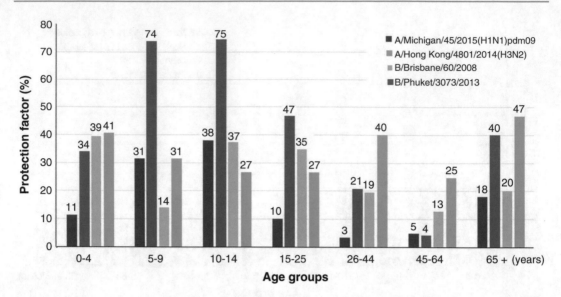

Fig. 2 Protection factor resulting from the level of antihemagglutinin antibodies in successive age groups in the influenza epidemic season 2017/18 in Poland

age belong to a high risk population for influenza infection, protection factor in the 2017/2018 epidemic season was clearly higher in the age groups up to 14 years, than in the adult groups older than 15 years of age. Of note, in case of influenza B virus lineages, Yamagata lineage enjoyed a higher protection factor than that of Victoria lineage in the current season, which was due likely to the predominance of Yamagata lineage as opposed to the previous season when Victoria linage predominated (Flu News Europe 2018). However, generally low levels of protection factor, lower than the set safety limit of 70% (Brydak 2008) outstandingly point to, and result from, a very low percentage of population being vaccinated against influenza in Poland. On average, this vaccination rate amounts to 3.6% in the current season. In the age group of 0–4 years it was 0.60% in 2016/17, which dropped to 0.54% in 2017/18 season, with a slightly better result in the group of 5–14 years where it was about constant around 0.94% in both seasons. A low percentage of the vaccinated children population up to 14 years of age indicates that the protective antihemagglutinin antibody titer is a result of antecedent infection.

Acknowledgments Funded by NIPH-NIH thematic subject 4/EM. We would like to acknowledge physicians and employees of the Voivodship Sanitary Epidemiological Stations across the country who collected sera and prepared epidemiological data. We also thank the Malec Company for providing fertilized chicken embryos.

Conflicts of Interest The authors declare no conflicts of interest in relation to this article.

Ethical Approval All procedures performed in studies involving human participants were in accordance with the ethical standards of the institutional and/or national research committee and with the 1964 Helsinki declaration and its later amendments or comparable ethical standards. The study was approved by an institutional Ethics Committee.

Informed Consent Informed consent was obtained from all individual participants included in the study before collection of nasopharyngeal samples.

References

ACIP (2016) Advisory committee on immunization practices. Centers for Disease Control and Prevntion (CDC). Prevention and control of seasonal influenza with vaccines: recommendations of the Advisory Committee on Immunization Practices – United States, 2016–17 influenza season. MMWR 65(5):1–54

Brydak LB (2008) Influenza, pandemic flu, myth or real threat? Rytm, Warsaw (in Polish)

Brydak LB, Machała M, Myśliwska J, Myśliwski A, Trzonkowski P (2003) Immune response to influenza vaccination in an elderly population. J Clin Immunol 23(3):214–212

Flu News Europe (2018) Joint ECDC–WHO/Europe weekly influenza update. https://flunewseurope.org/archives/viruscharacteristics. Accessed on 24 June 2019

Hallmann-Szelińska E, Cieślak K, Kowalczyk D, Szymański K, Brydak LB (2018) Antibodies to influenza virus hemagglutinin in the 2016/2017 epidemic season in Poland. Adv Exp Med Biol 1108:69–74

Kowalczyk D, Szymański K, Cieślak K, Hallmann-Szelińska E, Brydak LB (2019) Circulation of influenza virus in the 2015/2016 epidemic season in Poland: serological evaluation of anti-hemagglutinin antibodies. Adv Exp Med Biol 1150:77–82

NIPH-NIH (2018) Epimeld. Influenza and influenza-like illness. http://wwwold.pzh.gov.pl/oldpage/epimeld/grypa/index.htm. Accessed on 28 Sept 2018

Pedersen JC (2014) Hemagglutination-inhibition assay for influenza virus subtype identification and the detection and quantitation of serum antibodies to influenza virus. Methods Mol Biol 1161:11–25

WHO (2011) Global influenza surveillance network. In: Manual for the laboratory diagnosis and virological surveillance of influenza. WHO Press, World Health Organization, Geneva. https://apps.who.int/iris/bitstream/handle/10665/44518/9789241548090_eng.pdf. Accessed on 29 Sept 2018

WHO (2017) Recommended composition of influenza virus vaccines for use in the 2017–2018 northern hemisphere influenza season. https://www.who.int/influenza/vaccines/virus/recommendations/2017_18_north/en/. Accessed on 28 Sept 2018

Advs Exp. Medicine, Biology - Neuroscience and Respiration (2019) 44: 75–80
https://doi.org/10.1007/5584_2019_437
© Springer Nature Switzerland AG 2019
Published online: 27 September 2019

Respiratory Virus Infections in People Over 14 Years of Age in Poland in the Epidemic Season of 2017/18

K. Szymański, K. Łuniewska, E. Hallmann-Szelińska,
D. Kowalczyk, R. Sałamatin, A. Masny, and L. B. Brydak

Abstract

People most at risk of influenza complications are the elderly with impaired immunity. Clinical picture of influenza virus infection includes symptoms such as chills, increased body temperature, dry cough, chest pain, or dizziness as well as headaches and muscle aches. In the diagnosis of influenza, quick and effective tests are necessary. Sensitive diagnostic methods of molecular biology require more time, but the result firmly confirm or exclude the presence of the genetic material of influenza or other respiratory viruses. Influenza vaccination plays an important role in combating influenza infection. Unfortunately, the awareness of vaccination benefits is insufficient in Poland. In this study we demonstrate the results of examination of 4,507 people aged over 14 years toward the influenza infection in the epidemic season of 2017/18. Most of the confirmed infections were reported in older people aged over 65, a high-risk population group. A low percentage of the vaccinated population may affect an increased number of confirmed influenza viruses in the elderly. The findings demonstrate a need to increase awareness of vaccination benefits, which is particularly essential to avoid influenza infection in the elderly.

Keywords

lImmunity · Elderly · Epidemic season · Influenza · Respiratory infection · Respiratory virus vaccination

K. Szymański (✉), K. Łuniewska, E. Hallmann-Szelińska, D. Kowalczyk, A. Masny, and L. B. Brydak
Department of Influenza Research, National Influenza Center, National Institute of Public Health – National Institute of Hygiene, Warsaw, Poland
e-mail: kszymanski@pzh.gov.pl

R. Sałamatin
Department of General Biology and Parasitology, Medical University of Warsaw, Warsaw, Poland

1 Introduction

There are four types of influenza viruses: A, B, C, and D. Types A and B are responsible for seasonal outbreaks. Type A viruses are divided into subtypes, and currently, there are A/H1N1/pdm09 and A/H3N2/ subtypes circulating in the population, while influenza B viruses belong to two lineages: B/Victoria and B/Yamagata. At present, only type A viruses are responsible for causing a pandemic flu (CDC 2018a; Ghebrehewet et al. 2016). The population groups most vulnerable to complications associated with influenza virus infections are the elderly, children under 5 years of age, people with chronic diseases, pregnant women, health care workers, and people with immunological diseases that impair natural immunity (WHO 2018).

General symptoms of influenza virus infection consist of a sudden appearance of high fever, cough, headache, myalgia, and general malaise. Respiratory symptoms include sore throat and dry painful cough. In addition, symptoms from other organ systems may appear such as anorexia, dizziness, muscle pain, and nausea and vomiting in children. These symptoms would appear after a short incubation period of 1–2 days (ECDC 2018; Paules and Subbarao 2017).

An effective laboratory diagnosis of influenza is necessary to promptly provide appropriate antiviral treatment of infection. The fastest diagnostic method is a rapid influenza swab-based testing that shows sensitivity of about 70% or between 27% and 61%, depending on the source. The advantage of this test the speed as it takes 15–30 min to complete and the ease of performing it as it can be done outside a specialized laboratory. The disadvantage, however, is a substantial risk of false negative results due to rather low sensitivity or false positive results due to misinterpretation of weak bands in the test (Dwyer et al. 2006). The reverse transcription-polymerase chain reaction (RT-PCR) remains the most sensitive method for the detection of viral genetic material. This method takes less than a working day to complete a test and the advantage is the ability to detect more than just one respiratory virus when a multiplex set is used. Another possible method is the immunofluorescence-based reaction of hemagglutination inhibition, which defines the antiviral antibody titers (Kissova et al. 2014).

Protective vaccination is important in combating the influenza virus. Studies have shown that people vaccinated against influenza have significantly fewer influenza-like virus infections (Taksler et al. 2015). The knowledge about the benefits of influenza vaccination is insufficient in Poland, among both the medical personnel and the lay public. Parents do not vaccinate their children against influenza due mostly to the lack of knowledge about the benefits of it (Wozniak-Kosek et al. 2015). The risk–benefit ratio of vaccination has changed downward over the past decades, reducing mortality, particularly among adults (Pfleiderer 2014; Brydak 2012). A greater coverage rate of vaccination also substantially reduces the economic impact of influenza infection. A previous study showed that about 50% of older people who had been vaccinated developed influenza in the season of 2015/16, but they did not require hospitalization, which points to a mild course of the infection. Influenza vaccine also is fully effective in older people with the underlying chronic pathological conditions (Rondy et al. 2017). The present study seeks to determine the incidence of influenza infection in persons aged over 14 years in the epidemic season of 2017/18 in Poland.

2   Methods

2.1   Patients and Samples

In this study we examined over 4507 specimens obtained from patients over 14 years of age during the epidemic influenza season of 2017/18 in Poland. The specimens consisted of throat and nasal swabs or of bronchial tree lavage. The patients were divided into the following age groups: 15–25, 26–44, 45–64, and 65+ years of age. Genetic material of respiratory viruses was isolated from the specimens, using 200 μL samples suspended in sterile phosphate-buffered saline solution. Viral RNA was isolated with the Maxwell 16 Viral Total Nucleic Acid Purification Kit (Promega Corporation; Madison, WI). The procedure was carried out in accordance with the manufacturer's instructions.

2.2   Molecular Procedures

Real-Time Polymerase Chain Reaction (PCR)
The primers and probes were obtained from the International Reagent Resource run by the Centers for Disease Control and Prevention (CDC) in the US. Briefly, RNA was reversely transcribed at 50 °C for 30 min. The DNA obtained was subjected to the initial denaturation process (1 cycle at 95 °C for 2 min), followed by 45 cycles of amplification consisting of denaturation at 95 °C for 15 s, annealing at 55 °C for 10 s,

and elongation at 72 °C for 20 s. Positive control was the viral RNA obtained from the vaccine strains for the epidemic season of 2017/18 (A/Michigan/45/2015 pdm09-like virus, A/HongKong/4801/2014, and B/Brisbane/60/2008), and negative control was RNase-free water. The PCR reactions were performed using the LightCycler 2.0 System (Roche Diagnostics; Rotkreuz, Switzerland).

Conventional Multiplex PCR The tests were performed to identify the presence of the following respiratory viruses in the sampled material: influenza A and B; adenovirus; respiratory syncytial virus (RSV) A and B; human metapneumovirus (hMPV); human coronavirus (hCoV); human parainfluenza viruses 1, 2, 3; human bocavirus; and enterovirus. The RV15 OneStep ACE Detection Kit (Seegene; Seoul, South Korea) was used to determine the infection by the influenza-like viruses above outlined.

3   Results and Discussion

Distribution of the investigated specimens by age of patients is presented in Fig. 1. The patients aged 65+ constituted the most numerous group, followed by the groups of 45–64 and 26–44 years of age, with the least numerous being 15–25 years group. The prevailing type

of infection was influenza B virus, followed by type A unsubtyped and the A/H1N1/pdm09 and A/H3N2/ subtypes (Fig. 2).

Concerning the prevailing influenza type B infections, the greatest incidence was reported in persons aged 65+, followed by the groups of 15–25, 45–64, and 26–44 years of age. The majority of infections with type A virus unsubtyped were recorded in the group of 26–44 years of age, followed by the groups of 45–64 and 65+ years of age, with the least frequent in the 15–25 years group. The A/H1N1/pdm09 subtype was noted, in decreasing order of frequency, in the groups of 26–44, 45–64, 15–25, and 65+ years of age. The prevalence of infections with the A/H3N2/ subtype did not exceed 1.5% in any of the age groups. This subtype was the most prevalent pathogen in the group of 15–25 years of age, with a decreasing incidence in the successive age groups to reach a barely noticeable level of 0.29% in the elderly (Fig. 3).

In contradistinction to influenza infections, influenza-like infections with respiratory viruses were most often diagnosed in the youngest adult group of 15–25 years of age. One half of these infections were due to RSV, followed by PIV-1 and 2, hCoV, and hMPV. In the group of 26–44 years of age, a similar number of respiratory viruses was detected, with RSV, PIV-2 and 3, and rhinovirus being the most frequently

Fig. 1 Percentage distribution of investigated specimens by patients' age during the influenza epidemic season of 2017/18 in Poland

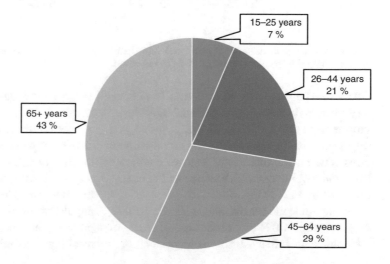

15–25 years
7 %

26–44 years
21 %

65+ years
43 %

45–64 years
29 %

Fig. 2 Prevalence of confirmed infections with influenza viruses in patients over 14 years of age during the epidemic season of 2017/18 in Poland

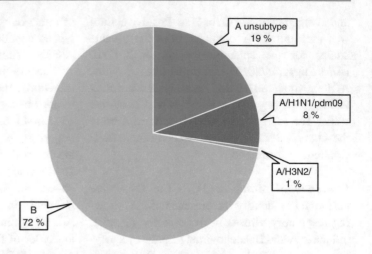

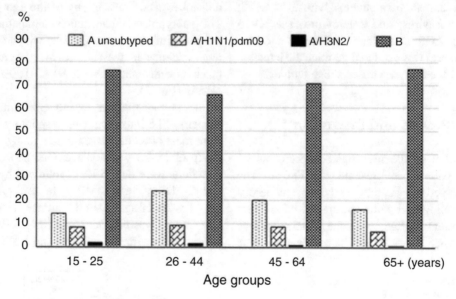

Fig. 3 Percentage distribution of confirmed infections with influenza viruses and their subtypes by successive age groups during the epidemic season of 2017/18 in Poland

detected. The frequency of detection of respiratory viruses decreased with patients' age. Fewer viruses were detected in the 45–64 age group, and they were identified as RSV, PIV-1, and coronavirus. The most seldom infections were reported in the elderly, in whom RSV was most frequently, followed by PIV-1 (Fig. 4).

In the epidemic season of 2017/18 in Poland, there were 2,971,031 cases and suspected cases of infection with influenza viruses in persons over 14 years of age, of which 2,455,933 cases occurred in people aged 15–64 years and 515,098 in those over 65 years of age. The percentage of people vaccinated against influenza in the fall of 2017 was 1.57% and 7.52% in the respective cohorts. For comparison, the vaccination rate was 3.7% in the 2017/18 season in the population of the entire country (PZH 2018). To protect the population from complications resulting from influenza infection, it is paramount

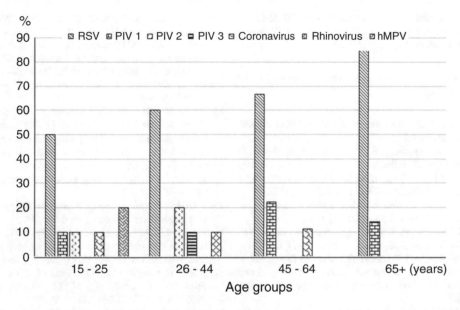

Fig. 4 Prevalence of confirmed infections with respiratory viruses in patients over 14 years of age during the epidemic season of 2017/18 in Poland. *RSV* respiratory syncytial virus, *PIV* parainfluenza virus, *hMPV* human metapneumovirus

to increase the vaccination coverage, mainly in people at risk of infection. The greater the number of people inoculated in their environment, the lower is the general risk of infection. Thus, those who cannot be vaccinated against influenza, for instance, due to allergies to vaccine ingredients, such as gelatin or some antibiotics, which often concerns children below the age of 6 months, would also be protected to an extent on the basis of collective resistance (CDC 2018b). Previous studies have shown that the vaccination coverage in a population should range from 33% to 73% to maximize the benefits of collective immunity and to reduce the transmission of influenza viruses (Logan et al. 2018; Plans-Rubió 2012). The present report demonstrates that influenza viruses can infect an adult person of any age. A study by Caini et al. (2018) have pointed, however, to the possibility of some age-dependent patterns of infection in that influenza B is more frequently detected in older children, while the subtype A(H3N2) is more common in the elderly.

In the 2017/18 influenza epidemic season, a larger number of specimens taken from people above 14 years of age were analyzed compared to the past 2016/17 season; 4507 and 2882, respectively (Kowalczyk et al. 2018). The distribution of infections among the age groups was, however, similar in the two seasons, with the larger number of positive results recorded in people over 65 years of age and a decreasing number of infections with decreasing age, with the smallest number in the 15–25 years group. In conclusion, the findings of this report emphasize a need to increase awareness of vaccination benefits, which is particularly essential to avoid influenza infection in the elderly.

Acknowledgments Supported by NIP-NIH thematic subject 3/EM.1. The authors express thanks to the physicians and employees of VSESs in the SENTINEL program for their input in the influenza surveillance in Poland.

Conflicts of Interest The authors declare no conflict of interests in relation to this article.

Ethical Approval All procedures performed in studies involving human participants were in accordance with the ethical standards of the institutional and/or national research committee and with the 1964 Helsinki declaration and its later amendments or comparable ethical standards. The study was approved by an institutional Ethics Committee.

Informed Consent Informed consent was obtained from all individual participants included in the study before collection of nasopharyngeal samples.

References

Brydak LB (2012) Influenza – an age old problem. Hygeia Public Health 47(1):1–7

Caini S, Spreeuwenberg P, Kusznierz GF et al (2018) Distribution of influenza virus types by age using case-based global surveillance data from twenty-nine countries, 1999-2014. BMC Infect Dis 18(1):269

CDC (2018a). https://www.cdc.gov/flu/about/viruses/types.htm. Accessed on 4 Oct 2018

CDC (2018b). https://www.cdc.gov/flu/professionals/vaccination/vaccine_safety.htm. Accessed on 4 Oct 2018

Dwyer DE, Smith DW, Catton MG, Barr IG (2006) Laboratory diagnosis of human seasonal and pandemic influenza virus infection. Med J Aust 185(10 Suppl): S48–S53

ECDC (2018). https://ecdc.europa.eu/en/seasonal-influenza/facts/factsheet. Accessed on 17 Aug 2018

Ghebrehewet S, MacPherson P, Ho A (2016) Influenza. BMJ 355:i6258

Kissova R, Svitok M, Klement C, Madarova L (2014) Factors affecting the success of influenza laboratory diagnosis. Cent Eur Public Health 22(3):164–169

Kowalczyk D, Szymański K, Cieślak K, Hallmann-Szelińska E, Brydak LB (2018) Respiratory infections with particular emphasis on influenza virus activity in persons over 14 years of age in the epidemic season 2016/2017 in Poland. Adv Exp Med Biol 1108:75–80

Logan J, Nederhoff D, Koch B, Griffith B, Wolfson J, Awan FA, Basta NE (2018) What have you HEARD about the HERD? Does education about local influenza vaccination coverage and herd immunity affect willingness to vaccinate? Vaccine 36(28):4118–4125

Paules C, Subbarao K (2017) Influenza. Lancet 390 (10095):697–708

Pfleiderer M, Trouvin JH, Brasseur D, Gränstrom M, Shivji R, Mura M, Cavaleri M (2014) Summary of knowledge gaps related to quality and efficacy of current influenza vaccines. Vaccine 32(35):4586–4591

Plans-Rubió P (2012) The vaccination coverage required to establish herd immunity against influenza viruses. Prev Med 55(1):72–77

PZH (2018). http://wwwold.pzh.gov.pl/oldpage/epimeld/grypa/index.htm. Accessed on 4 Oct 2018

Rondy M, Larrauri A, Casado I et al (2017) 2015/16 seasonal vaccine effectiveness against hospitalisation with influenza A(H1N1)pdm09 and B among elderly people in Europe: results from the I-MOVE+ project. Euro Surveill 22(30):pii: 30580. https://doi.org/10.2807/1560-7917.ES.2017.22.30.30580

Taksler GB, Rothberg MB, Cutler DM (2015) Association of influenza vaccination coverage in younger adults with influenza-related illness in the elderly. Clin Infect Dis 61(10):1495–1503

WHO (2018). http://www.who.int/en/news-room/factsheets/detail/influenza-(seasonal). Accessed on 17 Aug 2018

Wozniak-Kosek A, Mendrycka M, Saracen A, Kosek J, Hallmann-Szelińska E, Zielnik-Jurkiewicz B, Kempińska-Mirosławska B (2015) Vaccination status and perception of influenza vaccination in the polish population. Adv Exp Med Biol 836:41–46

Advs Exp. Medicine, Biology - Neuroscience and Respiration (2019) 44: 81
https://doi.org/10.1007/978-3-030-34651-5
© Springer Nature Switzerland AG 2019

Index

Printed in the United States
By Bookmasters